Guide to the Medical Evaluation for Respirator Use

ROBERT K. MCLELLAN, MD, MPH
KATHLEEN M. SCHUSLER, MSN, COHN-S

OEM PRESS

OEM Press
Beverly Farms, MA

For Lucy, Quinby, and Eliza—
They believed me when I said there would be a day when this book would be done.

RKM

For Steven
What is all this sweet work worth . . . PB Shelley

KS

ISBN: 1-883595-30-4

Printed in the United States of America

Published by OEM Press
8 West Street
Beverly Farms, Massachusetts
1-800-533-8046

CONTENTS

Preface v

Introduction ix

1. The Rationale for Medical Evaluation for Respirator Use 1
2. Before the Medical Evaluation: An Overview of the Employer's Responsibilities 27
3. Medical Evaluation Procedures 49
4. Medical Decision Making for Respirator Approval 75
5. Special Problems with Respirator Use 185
6. Communicating Respirator Use Recommendations 193
7. Building Continuous Quality Improvement into the Medical Evaluation for Respirator Use 213

Appendices

Appendix 1. OSHA Respiratory Protection Standard 29CFR 1910.134 219

- Appendix 1A. Appendix A to § 1910.134 (Mandatory): Fit Testing Procedures 241
- Appendix 1B1. Appendix B1 to § 1910.134 (Mandatory): User Seal Check Procedures 263
- Appendix 1B2. Appendix B2 to § 1910.134 (Mandatory): Respirator Cleaning Procedures 264
- Appendix 1C. Appendix C to § 1910.134 (Mandatory): OSHA Respirator Medical Evaluation Questionnaire (English) 266
- Appéndice 1C. Cuestionario de Evaluación Médico obligado por la OSHA (La agencia de seguridad y salud ocupacional) Parte 29 CFR 1910.134 Mandatorio para Proteccion del Sistema Respiratorio (Español) 275

Appendix 1D. Appendix D to § 1910.134 (Mandatory): Information for Employees Using Respirators When Not Required Under the Standard 284

Appendix 2. Environmental Conditions and Respirator Use 285

Appendix 3. Physical Examination Form: Respirator Use Qualification 287

Appendix 4. Consent and Release Form 289

Appendix 5. Spirometry Testing as Part of the Respiratory Certification Process 291

Appendix 6. Olfactory Acuity Testing Procedure 301

Appendix 7. Work Trial Procedure 305

Index **319**

PREFACE

Several years ago we developed a quality improvement program for our community hospital-based occupational health clinic. The hospital's quality improvement team suggested that we focus on activities that were either high risk or high volume. In contrast to emergency medicine, in-patient clinical activities, or even primary care, occupational health care seldom involves risky procedures. Instead, much of occupational clinical practice involves high-volume examinations of generally healthy people. On behalf of employers, the practitioner conducts these examinations for regulatory compliance, medical surveillance, and to make recommendations for job placement and respirator use.

Occupational health practitioners distinguish themselves from their colleagues in other disciplines by their better understanding of the medical aspects of the relationship between workers and their jobs. This unique knowledge prepares the occupational health professional to rationally advise employers and employees about the capacity of employees to perform specific job tasks safely. Therefore, we chose to review indicators of the quality of our preplacement and respirator certification recommendations. We soon learned how difficult this task would be.

The American College of Occupational and Environmental Medicine (ACOEM) considers the ability to perform respirator certification examinations a core competency.[1] In 1990, a brief review of standard occupational medicine and nursing textbooks revealed an absence of *any* guidelines helpful in evaluating a worker for respirator wear. Referring to the existing Occupational Safety and Health Administration (OSHA) respirator standard was equally futile. While requiring a medical evaluation, the standard left the content of the examination entirely up to the individual practitioner. Further research revealed an American National Standards Institute (ANSI) standard that provided a suggested evaluation protocol. It also provided an outline of medical conditions or physiologic states that might preclude the use of a respirator. A few calls to corporate medical directors revealed that some companies have guidelines and medical standards for respirator wear.

A literature search revealed a handful of articles that detailed a suggested evaluation protocol for making respirator use recommendations. Beyond face validity, however, we could find scant information that supported evidence-based respiratory qualification decisions. In brief, little guidance was easily available to the front-line occupational health practitioner interested in making evidence-based medical recommendations for respirator use.

Recognizing that we could not readily assess the quality of respirator use recommendations themselves, we resorted to structural indicators of quality. We decided that clinicians must have certain data to make a good recommendation: They must know what respirator will be used, what job demands will be placed on the respirator user, and some basic clinical data. We quickly recognized that few of our clinic's charts had *any* information about job functions or respirators. Standard brief examinations were being conducted, but pulmonary function testing results revealed technician deficiencies.

We proceeded to attempt to correct these egregious shortcomings. Technicians were provided additional training, improved protocols, and quality review. We developed a simple form that was given to all employers sending workers to our clinic for evaluation for respirator use. The form asked basic information about job demands and the types of respirators to be used. Employer compliance remained poor. Examiners continued to make respirator use recommendations without adequate information.

Times have changed. In 1996, Phil Harber broke new ground by including a section on respirator evaluation in his textbook entitled, *Occupational and Environmental Respiratory Disease*. Since then, other classic occupational medicine textbooks and articles in the *Journal of Occupational and Environmental Medicine* have provided some practical guidance to clinicians making decisions about approving workers for respirators. The Americans with Disabilities Act of 1992 clearly calls in question traditional respirator use recommendations based on conventional rather than evidence-based practice. Disqualifying a worker with a disability for respirator use without good medical reason is unacceptable. More science, albeit not much, has become available to inform decisions.

Most significantly, in 1998, after long deliberations and substantial public input, OSHA promulgated a new respiratory standard that has redrawn the landscape. Several features of this standard highlight a new standard of care to be used in respirator use medical decision making. Employers are required to provide the health care examiner with information about the respirators used and the examinee's job demands. A mandated questionnaire can be used in lieu of a physical examination. The standard authorizes licensed health care practitioners (LHCPs) other than physicians to make medical recommendations for respirator wear. Controversy erupted as ACOEM questioned the wisdom of the decision to permit LHCPs to make respirator use recommendations without the supervision of a physician. OSHA and the courts have turned down ACOEM's request for reconsideration of this position. On August 4, 1999, the National Institute of Safety and Health (NIOSH) issued a policy statement that endorsed all provisions of OSHA's new respiratory protection standard with the exception of the use of irritant smoke for qualitative respirator fit testing and "unsupervised medical evaluations conducted by health care professionals who are not licensed for independent practice to perform or supervise medical evaluations."[2] However one might feel about the appropriate scope of practice for LHCPs, all can agree that physicians and LHCPs must interact effectively to optimize efficiencies and the quality of decision making. Unfortunately, the standard does not provide the clinical protocols necessary to achieve this goal.

We have written this book to redress what we feel has been a glaring lack of hands-on guidance for one of the most common occupational health activities—medical evaluation for respirator use. Although we remain dismayed at the paucity of data available for truly evidence-based decisions, we hope to advance clinical practice.

Acknowledgments

To bring this project to fruition, we depended on the encouragement, critique, support, and patience of many people. Good questions from our colleagues at the Center for Occupational and Environmental Health provided the inspiration to start this book. Curtis Vouwie, our publisher, trusted that we could do it. Marty Tenney, project manager, somehow read our scribbled edits to the original manuscript and got this project to the printer on time. Phil Harber taught us more than anyone about respirators and worker health. Ronald Mack provided invaluable, authoritative criticism. Mary Townsend lavished us with time and spirometry expertise. Nurse practitioners Trudi Buckley and Kathleen Golden McAndrew made sure we made sense. North Safety Products and MSA generously provided pictures of respirators workers. And, Alice Reed, librarian extraordinaire, found every article we asked for.

References

1. American College of Occupational Environmental Medicine Panel to Define the Competencies of Occupational Environmental Medicine. Occupational environmental medicine competencies, version 1.0. *J Occup Environ Med* 1998;40:427–440.
2. Rosenstock L. *NIOSH Respirator Use Policy.* Washington: NIOSH, August 4, 1999.

INTRODUCTION

This book provides the busy occupational health practitioner with medical guidelines for the rational evaluation of workers for respirator use. The text contains all the information necessary for setting up, conducting, and auditing a program for the medical evaluation of respirator users. Protocols describe how physicians and other licensed health care providers can work together to maximize both the efficiency and quality of decision making. We do not cover all the aspects of a respiratory protection program or the Occupational Safety and Health Administration (OSHA) respirator standard. Our focus is on the role of the occupational health clinicians involved in the medical evaluation for respirator use.

The new OSHA respirator standard permits both physicians and licensed health care professionals (LHCPs) to provide employers with recommendations regarding an individual's use of a respirator. This book will help all occupational health practitioners perform this duty. Although targeted at clinicians, other health and safety professionals with a responsibility or interest in respiratory protection will find this book useful. Lawyers and human resources professionals will find the sections on accommodating workers with medical problems helpful in executing their duties.

What This Book Is About

This book helps physicians and LHCPs competently and efficiently discharge their responsibilities as directed by OSHA's revised respiratory protection standard. Some occupational health professionals may argue that respirator qualification involves only the decision as to whether an individual can wear a respirator, without reference to the job demands for which that respirator will be used. We could not disagree more! Because employees use respirators to work, a medical opinion that disregards the actual demands that will be placed on a respirator user falls inexcusably short of providing useful advice. Such advice parallels a recommendation made by a work-capacity examiner that merely states that an individual can work, without detailing what this person can and cannot do. OSHA's new requirement that employers provide the physician or LHCP with information about job demands and the work environment establishes the regulatory framework for this more expansive view of the medical evaluation for respirator use. At its heart, this book guides physicians and LHCPs to incorporate answers to the following questions in their recommendations to employers about an individual's use of a respirator:

1. Can the individual wear a specific type of respirator safely while performing a particular task in the work environment?
2. Will the worker comply with respirator use?
3. What may be the consequences of respirator failure?

The reader will find:

- Comprehensive guidelines for compliance with the OSHA respirator standard for medical evaluation
- Employer responsibilities for optimizing the medical evaluation
- A description of respirators and the demands they place on workers
- Decision-making logic for nurses and physicians
- Scientific evidence for medical decision making
- Approaches to accommodating workers with health problems that may interfere with safe use of a respirator
- Approaches for continuously improving the medical evaluation program
- Appendices that gather together in one place all relevant regulations, forms, and procedures

The first chapter defines the rationale and intent of medical evaluations for respirator use. Employers may ask their employees to wear any of a variety of respirators. An examiner cannot rationally approve an individual for respirator use without a good understanding of the respirator he or she will be using. The differing characteristics of respiratory protection devices are explained, with an emphasis on the physiologic and psychological demands they place on workers. This chapter also presents the scientific basis for integrating job-task analysis with knowledge of the stresses of respirator use to optimize the quality of the medical assessment.

Chapter 2 outlines the employer's responsibilities for respiratory protection mandated by the OSHA standard. The text leads the reader systematically through the tasks necessary to comply with the standard. The employer must attend to multiple factors besides medical fitness to ensure the success of a respiratory protection program. We emphasize the features of the standard that pertain to the medical evaluation. Occupational health practitioners cannot do their job in a vacuum. Clinicians need employer input regarding the types of respirators an employee may wear, job demands while wearing the respirator, and information about any other personal protective equipment that the employee may be wearing.

Chapter 3 provides the details for choosing and implementing a procedure of medical evaluation. This chapter offers specific recommendations about the administration of the questionnaire and the contents of the medical examination. Employers choosing a medical evaluation method will want to focus on a review of the pros and cons of each approach. We define the nature of a follow-up evaluation and tests commonly conducted at this stage, with special reference to protocols for spirometry, olfactory acuity testing, and work trials.

Chapter 4 provides the medical foundation for decision making regarding respirator use. Referencing questions from the OSHA questionnaire and ordered by organ system, this chapter describes how various diseases and psychological problems may affect a worker's ability to tolerate a respirator. Detailed guidelines offer physicians and LHCPs advice on respirator approval for workers with relevant medical conditions. The advice proceeds from simple measures that LHCPs can take, without additional consultation, to essential considerations useful to the advanced practitioner's decision making. Whenever possible, we cite scientific evidence, regulatory guidance, or medical organizations that support the recommendations. We provide suggestions about education and accommodations for workers with health problems that may interfere with respirator tolerance.

Chapter 5 addresses special circumstances that may prompt an employer to request medical recommendations from a physician or a LHCP. To begin, we provide guidance regarding problems with respirator fit. Physicians and LHCPs may prescribe respirator use as an accommodation for individuals with enhanced susceptibility to airborne hazards at the workplace. Physicians and LHCPs confronted with a request to review the appropriateness of such a prescription will benefit from the protocol offered.

Chapter 6 provides guidance to physicians and LHCPs in formulating their respirator approval recommendations to employers.

We suggest language for writing restrictions for an employee's respirator use and provide several examples of recommendations for common health problems.

Chapter 7 offers occupational health managers the tools to create a continuing quality improvement system for their medical evaluation for respirator use program.

The appendices collect a comprehensive resource of regulations, forms, and procedures essential to implement a program for the medical evaluation of respirator users.

How to Use This Book

We have written a manual for clinical practice, not a textbook. We hope that it is quickly dog-eared and worn. Readers should familiarize themselves with the entire contents of Chapters 1, 2, 3, 6, and 7. The numbered guidelines in Chapters 4 and 5 contain redundancies. Physicians and LHCPs thus can access all the information necessary for decision making in terms of a particular medical problem without having to read the entire chapter. These two chapters are more appropriately used for reference as the need arises than read from beginning to end at one sitting. These chapters also offer a well-referenced narrative that reviews the scientific and regulatory basis for our recommendations for those readers interested in accessing the sources for our guidance. Practitioners may choose to bypass these sections and focus only on those parts of the book which provide the tools to implement the medical evaluation process. We furnish procedures and other clinical information in easily digested practical parcels. Whenever appropriate, we offer sample forms that can be copied.

1

The Rationale for Medical Evaluation for Respirator Use

The sheer volume of respirator use demands the attention of occupational health professionals and employers. The Occupational Safety and Health Administration (OSHA) estimates that about 5 million workers use respirators.[1] They use them for respiratory protection from a variety of occupational exposures. The spectrum of hazards requires that workers use the correct respirator to protect themselves. Each type of respirator places unique physiologic and psychological demands on a worker. Almost every industry uses respirators to some extent. OSHA believes that 1.3 million businesses, or 20 percent of all businesses, have employees who use respirators.[1] In 1971, OSHA based its first respiratory protection standard (29 CFR 1910.134) on a voluntary consensus standard. Since then, the OSHA regulation has prompted millions of medical evaluations for respirator use, a large percentage of all work-related examinations. Thirty-nine percent of all companies with medical surveillance programs conduct them for purposes of assessing workers for use of personal protective equipment.[2]

This chapter delineates the scientific rationale for the medical assessment of personnel wearing respirators. First, a distinction is drawn between respiratory medical surveillance and medical evaluation for respirator use. The next section describes the features of respirators and their associated psychophysiologic stresses. Individuals do not simply wear respirators. Workers must perform their assigned job tasks while wearing their protective devices. We describe the interaction between job demands and respirator use.

Distinguishing Medical Evaluation for Respirator Use from Respiratory Surveillance

Many people confuse medical evaluation for respirator use with respiratory health surveillance. In each case, the medical examiner evaluates the worker with the goal of preventing occupational disease. Overlap between the medical content of the two programs helps confound their differences. Moreover, workers may participate in both programs. To make matters yet more complex, in the past, several OSHA health standards have addressed both respirator use and respiratory health surveillance. To fulfill their responsibilities, physicians and licensed health care professionals (LHCPs) must understand the different goals and characteristics of these occupational health evaluations. Table 1-1 summarizes these chief distinctions.

Respiratory Health Surveillance

Respiratory health surveillance aims to prevent occupational disease. Accomplishing this goal depends on systematic medical data collection and analysis over time. Respiratory health surveillance assesses the effect that a specific hazard or occupation has on the respiratory system. Both the individuals screened and the populations of workers at risk for similar exposure may benefit from these data. In some cases, the information also may further scientific knowledge.

Respiratory health surveillance must occur periodically to identify early signs of respiratory disease that will trigger appropriate clinical and industrial hygiene interventions. Typically, such a program includes preplacement, interval, post-accidental exposure, and termination examinations. The expected natural course of illness created by a particular respiratory hazard determines the frequency of interval rescreening. Because of the long latency of asbestos-related diseases, rescreening is scheduled at long intervals. Chest x-rays, for example, are ordered only every 5 years for individuals whose first exposure to asbestos began less than 10 years ago. On the other hand, the short latency of isocyanate asthma dictates that surveillance should be scheduled at short intervals. After the preplacement evaluation, the first rescreening may occur at 6 weeks, with repeat screening at 3 months and perhaps 6 months.

Several OSHA health standards mandate respiratory surveillance. Examples include the asbestos, cotton dust, and formaldehyde standards. Many employers also have established surveillance programs not required by OSHA. Examples of employee populations in nonregulated employer-sponsored programs include isocyanate workers and firefighters. Pulmonary surveillance programs may include the collection of interview or questionnaire data about respiratory symptoms, physical examinations targeted on the respiratory system, chest x-rays, spirometry, and occasionally, other laboratory data.

Table 1-1. Distinctions Between Medical Evaluation for Respirator Use and Respiratory Health Surveillance

Characteristic	Evaluation for respiratory use	Respiratory health surveillance
Focus	Individual	Population
Scope	Respirator users	Workers exposed to respiratory hazards above OSHA-mandated or voluntary action levels
Goals	1. Prevent occupational disease 2. Assess fitness to wear respirator	1. Prevent occupational disease 2. Assess health effects due to respiratory hazards
Who benefits	Individual	Individual and population
Timeline	Focus on current health status	Comparison of health status over time
Triggers for re-evaluation	Individual health status or changes in job demands	Individual health status or characteristics of hazardous exposure
Significant health outcomes	Changes in clinical status of individual	Changes in population as well as individual clinical status
Data sources	Individual	Individual, databases
Regulation	29 CFR 1910.134	Multiple
Examiner	Physician or licensed health care professional	Physician or licensed health care professional supervised by the physician[a]

[a]Two exceptions to this general rule are the revised methylene chloride standard (29 CFR 1910.1052) and the recently promulgated tuberculosis standard (OSHA, Occupational exposure to tuberculosis: Proposed rule. *Federal Register* 62:54159–54309). ACOEM has filed suit to challenge OSHA's new approach to marginalizing physician supervision of medical surveillance.

Health-based respiratory surveillance helps prevent disease in populations of workers. The identification of illness in one individual serves to alert the professionals and regulators overseeing the program of a hazard that requires greater industrial hygiene control. Health outcomes not necessarily meaningful in one person nonetheless may identify a significant respiratory hazard. Population mean spirometric changes that are insignificant clinically nonetheless may signal the failure of industrial hygiene programs or even the discovery of a new respiratory hazard.

Unless conducted for research purposes only, surveillance examinations also assess an individual's medical capacity to work safely in a specific, potentially hazardous environment. As a tool for secondary prevention, surveillance helps establish the need for early intervention to prevent disease progression. The examination identifies individuals who may benefit from health care or from removal from further exposure to a specific hazard.

In addition to individual clinical review, occupational health professionals may use other approaches to conduct respiratory surveillance. They may design a passive surveillance system that uses data collected routinely for other purposes. A passive program may rely on such databases as workers' compensation loss runs, OSHA logs, or death certificates. A more inclusive definition of respiratory surveillance includes hazard surveillance as well. Exposure-based surveillance collects data about the work environment as a proxy for health outcomes. It is not helpful in identifying new hazards. Readers interested in furthering their knowledge of the surveillance and screening for occupational respiratory disease are referred to other resources listed at the end of the chapter.

Medical Evaluation for Respirator Use

Like respiratory health surveillance, medical evaluation for respirator use aims to prevent occupational disease. In contrast to surveillance, the examiner's single focus is protecting the health of the individual respirator user. The evaluation cannot depend on population databases. Most generally, this screening process strives to ensure that an individual will be able to wear a specific type of respirator safely while performing a particular task in a specific environment.

OSHA does not mandate time-based periodic reevaluations for respirator use. The physician or LHCP recommends both the content and periodicity of reevaluations for respirator use based on an individual's health status. The physician or LHCP considers such health indicators as age, change in health status, or change in job demands. Employers must refer an individual for reevaluation if they notice problems during respirator fit testing or while using a respirator. They also should request a reevaluation before asking an employee to wear a new type of respirator or perform tasks involving higher physical demands or different hazardous exposures.

Several OSHA standards contain mandates regarding respirator use. The new OSHA respiratory protection standard has simplified interpretation of the agency's directives by revising the respirator-related provisions of existing standards to achieve regulatory consistency. These revisions are compiled in the appendices to the respiratory protection rule. In contrast to most respiratory health surveillance standards, OSHA permits a simple questionnaire approach to medical evaluation. More in-depth assessment, including personal interviews, physical examinations, and laboratory testing, is required only if red flags are noted on the questionnaire. Moreover, LHCPs other than physicians may conduct the evaluation.

Respiratory Protection Devices

Any physician or LHCP conducting medical evaluations for respirator use must understand the characteristics of the various devices available for respiratory protection. Respirators differ in their methodology for protecting users, their protectiveness, their weight, and their fit. Each type of device places different demands on its user. The medical assessment for respirator use does not include fit testing, training, or the responsibility for choosing the correct respirator for a specific hazard. On the other hand, the physician or LHCP must understand that improper fit, incorrect use, or selection of the wrong respirator may increase a worker's exposure. In addition, the physician or LHCP does have the responsibility to recommend alternative respirators to accommodate workers with medical problems or facial deformities that may interfere with proper fit. Book knowledge of these devices fails to endow the physician or LHCP with a full appreciation of the stresses of wearing a respirator. Conscientious occupational health practitioners conduct work-site tours to better appreciate the job demands of the workers they care for. Similarly, physicians and LHCPs benefit from a trial of wearing each of the different types of respirators.

Respirator use presents a variety of occupational health and safety concerns. Few people enjoy wearing respiratory protection. At a minimum, respirators interfere with comfort, communication, vision, and movement. They place additional physical and psychological demands on workers in addition to those inherent in the job tasks. Together, personal comfort and medical considerations drive noncompliance, which can lead to hazardous exposures. Respirators can fail to protect human health for other reasons. Deficiencies in the overall respiratory protection program described in Chapter 2 and technological problems add to this failure rate. The physician or LHCP must always consider the consequences of respirator failure on the individual who is being assessed for respirator use.

Despite their problems, respirators remain a common and extremely important tool for worker protection whenever adequate environmental controls do not exist. They also may be used for protection from accidental releases, to allow workers to enter hazardous areas not typically occupied, or for safeguarding workers with special susceptibilities.

Types of Respirators

Employers must provide their employees with respirators certified by the National Institute of Occupational Safety and Health (NIOSH) to comply with 42 CFR Part 84. This respirator certification regulation was revised in 1995. It replaced 30 CFR Part 11, the respirator certification regulation under which NIOSH and the Mine Safety and Health Administration (MSHA) had jointly certified respirators since 1972.

Respirators protect workers by reducing their exposure to airborne contaminants. They do so either by removing contaminants from the ambient air (air-purifying), supplying breathing air from another source (atmosphere-supplying), or a combination of these methods. Respirators also may be characterized by their fit, pressure dynamics, respiratory inlet covering, and construction design (Table 1-2). Each of these factors affects worker comfort, protection, and physiologic and psychological stress.

The following sections emphasizes respirator features that are relevant to the medical evaluation. Readers interested in a more complete description of respirators are referred to the resources listed at the end of this chapter.

Air-Purifying Respirators

Air-purifying respirators represent the most commonly used respiratory protective devices. They employ filters, cartridges, or canisters to remove specific air contaminants by passing ambient air through the air-purifying element. Respiratory protection relies on the proper choice

Table 1-2. Characteristics of Respirators

- Exposure-reduction method
  - Air-purifying
    - Particulate removal
    - Vapor removal
    - Combination
  - Atmosphere-supplying
    - Self-contained breathing apparatus (SCBA)
      - Closed circuit (compressed oxygen, chemical oxygen, or liquid oxygen)
      - Open circuit (compressed air)
    - Supplied air
    - Combination
- Pressure dynamics
  - Negative pressure
    - Nonpowered air-purifying respirators
  - Positive pressure
    - Powered air purifying respirator (PAPR)*
  - Atmosphere-supplying
    - Continuous flow*
    - Demand (flow occurs with inspiration only)*
    - Pressure demand (flow occurs with inspiration, but device attempts to maintain positive pressure)
- Physical configuration
  - Air-purifying
    - Purification medium incorporated into mask
    - Cartridge
    - Canister
  - Atmosphere-supplying
    - Air-line
    - Air-line with backup tank
    - Self-contained breathing apparatus (SCBA)
- Respirator fit
  - Loose-fitting coverings
    - Hood
    - Helmet
  - Tight fitting facepieces
    - Full mask
    - Half mask
    - Quarter mask
    - Mouthpiece

*Deep inspiration may intermittently create negative pressure.
Source: Adapted from Harber P. Respirators. In: Rom WN (ed.), Environmental and Occupational Medicine, 3d ed. Philadelphia: Lippincott-Raven, 1998: 1757–1765.

of filters, sorbents, and catalysts specific to the type of exposure. Respirators frequently combine these methods to protect from complex exposures.

Filtration mechanisms rely on pore size and other physical factors to remove particles. NIOSH rates filtering respirators by the efficiency with which they remove various sizes of particles and whether oil-based aerosols degrade their filtration performance. In the revised 42 CFR Part 84, NIOSH provides three filter classifications, N, R, and P, with three levels of filter efficiency, 95, 99, and 99.97 percent, in each class. N-series filters are designed for use in workplaces free of oil or other severely degrading aerosols. R-series filters do not have aerosol-use restrictions but have other use limitations. The P-series filters have neither aerosol-use nor time-use limitations other than those based on filter loading.

Respirator manufacturers fill cartridges and canisters with a matrix impregnated with sorbents specific for the adsorption or absorption of a particular gaseous chemical agent. In addition to sorption, catalysts may be used to react with a contaminant to produce a less toxic gas or vapor.

Filters, sorbents, and catalysts must be replaced periodically to maintain their effectiveness. The service life of filtering respirators depends on hygiene considerations and increased breathing resistance due to filter loading. Historically, workers depended heavily on the odor and irritating properties of chemicals as the warning that the capacity of sorbents and catalysts had been exhausted. Now OSHA considers odor, taste, and irritation as unreliable or inappropriate to use as indicators of sorbent exhaustion. Instead, 29 CFR 1910.134 requires chemical cartridge end-of-service-life indicators (ESLIs) or change-out schedules for air-purifying respirators used for protection against gases and vapors. These regulations aim to ensure replacement of the sorbent at less than or equal to 90 percent of its service life.

Currently, ESLIs exist for few toxicants, mercury vapor, ethylene oxide, and hydrogen sulfide. These ESLIs rely on colorimetric indicators that do not require color vision for recognition of its warning. An electronic ESLI that provides a flashing light indicator has been tested and found to be feasible when used with a full-face mask.[3] Manufacturers can be expected to develop additional ESLIs in the future. In the meantime, manual tables and software are available to calculate change-out schedules.

The new OSHA requirement for ESLIs or change-out schedules reduces, but does not eliminate, the importance of a worker's sense of smell, taste, or irritation as a fail-safe method for detecting the need for changing a spent cartridge. Odor, taste, and irritant detection remain important, though imperfect, in identifying other causes of respirator failure, such as a faulty seal, and/or as a signal to don a respirator in unexpected hazardous conditions.

Respirator design may affect comfort, fit, and service life. Three basic configurations trade comfort for service life, since the more air-purification medium used, the heavier, more uncomfortable, and more awkward is the respirator. The lightest and most comfortable respirators are those which incorporate air-purifying elements into the mask itself (Fig. 1-1). These respirators are disposable and have the shortest service lives. Respirators also may employ cartridges or canisters that attach directly to the mask or connect to the facepiece by a hose. Cartridges contain more air-purification material than simple masks, and canisters contain even more (Figs. 1-2 and 1-3).

Powered air-purifying respirators (PAPRs) use a fan to force the ambient air through the purification medium to the respiratory inlet covering, relieving the user of this effort (Fig. 1-4). They are more bulky than negative-pressure respirators. Nonetheless, most workers find PAPRs more comfortable. They need not fit tightly against the face, offer some cooling, and require less ventilatory effort.

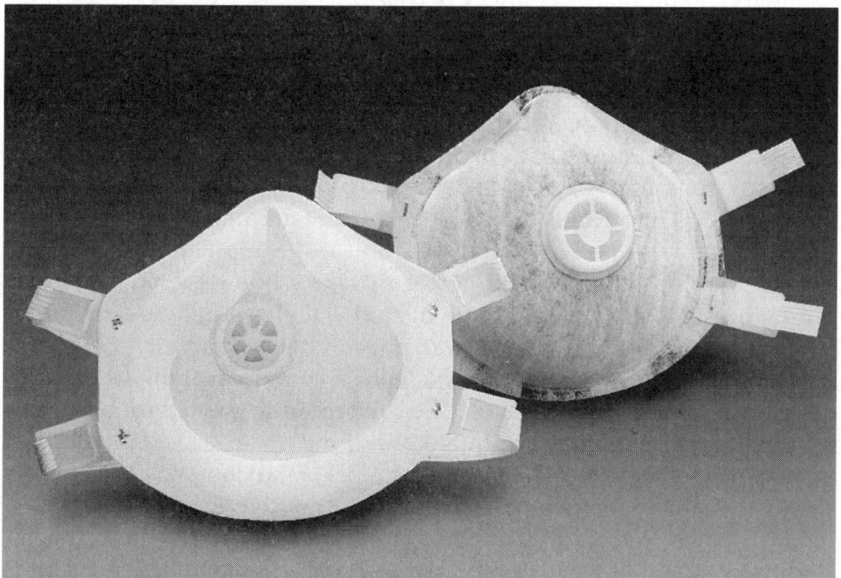

Figure 1-1. Disposable air-purifying respirator medium integrated with mask. (Reproduced courtesy of MSA.)

Figure 1-2. Half-mask negative-pressure air-purifying cartridge respirator. (Reproduced courtesy of MSA.)

Atmosphere-Supplying Respirators

Atmosphere-supplying respirators generally provide the highest levels of protection by supplying breathing air independent of the ambient work environment. Air may be delivered by a hose (air line or supplied-air respirator), by an air line with a backup air tank, or by a self-contained breathing apparatus (SCBA) (Figs. 1-5 and 1-6). Supplied-air respirators (SARs) require the worker to perform tasks while tethered to a hose, which presents a trip and tangle hazard. The hose delivers air under pressure from a compressor that must have a source of clean air. Manufacturers provide some SARs with a backup air tank in case the air line becomes inoperable. An SCBA provides greater mobility but requires workers to carry an air tank weighing up to 40 lb on their backs. SCBAs place greater aerobic and musculoskeletal demands on workers than any other respirator. Their weight and size also challenge the balance of workers as well as their ability to work in confined spaces. Lighter (7–10 lb), less bulky SCBAs are also available for brief use only, usually for escape purposes. SCBA use is also limited by the amount of air in the tank. The service time of the tank decreases as the aerobic demands of the work increase.

Figure 1-3. Air-purifying negative-pressure canister respirator. (Reproduced courtesy of MSA.)

Manufacturers offer SCBAs in two configurations: closed and open circuit. Closed-circuit devices recirculate air for rebreathing. They remove exhaled carbon dioxide and moisture and replenish the oxygen with liquid or compressed oxygen or a solid chemical oxygen generator such as potassium permanganate. Open-circuit devices exhaust exhaled air.

Respirator Pressure Dynamics

The pressure dynamics of a respirator greatly affect the protectiveness, ventilatory effort, thermal burden, and comfort of its user. A negative-pressure respirator requires that the user overcome the air resistance of

Figure 1-4. Powered air-purifying full-mask respirator. (Reproduced courtesy of MSA.)

the medium to inhale. As filters become loaded with particulates, their resistance to airflow increases. The negative pressure generated pulls the mask against the face. If the mask does not seal well on the face, contaminated air will bypass the respirator. Fit testing and the worker's routine field check decrease the likelihood of an improper seal. Powered air-purifying respirators (PAPRs) supply air at a positive pressure, but high ventilatory rates with heavy exertion may exceed flow rates and thereby generate negative pressures.

Atmosphere-supplying respirators may operate in one of three modes: continuous flow, airflow on inspiration (demand) only, or pressure demand. Continuous-flow respirators (not available in self-contained units, i.e., SCBAs) aim to maintain a positive pressure inside the face covering. As with PAPRs, however, a worker's ventilatory

Figure 1-5. Supplied-air full-face respirators. (*left*) Device includes backup tank. (Reproduced courtesy of MSA.)

requirement may overcome the flow rate, resulting in negative pressure. When flow through the respirator's intake valve depends on inspiratory demand, pressure also drops inside the mask. The potential for leakage again occurs. Leakage drops the protectiveness of an atmosphere-supplying respirator to the same level as a negative-pressure air-purifying respirator.[4] Pressure-demand respirators solve this problem by allowing actuation of the intake valve at pressures higher than the ambient air.

Respiratory Inlet Covering

OSHA defines a respirator inlet covering as the portion of a respirator that forms the protective barrier between the user's respiratory tract

Figure 1-6. Self-contained breathing apparatus. (Reproduced courtesy of North Safety Products.)

and the air-purifying device or breathing-air source, or both. It may be a facepiece, helmet, hood, suit, or mouthpiece respirator with a nose clamp. Most respirators rely on a mask that fits tightly, forming a complete seal on the face. A full-face mask extends from hairline to below the chin, helps protect the eyes, and provides the best seal (see Fig. 1-5). A quarter-mask covers the mouth and nose, with the lower surface resting between the chin and mouth. A half-mask extends to below the chin, allowing a more reliable seal (see Fig. 1-2). Escape respirators may employ only a mouthpiece. Atmosphere-supplying positive-pressure respirators and powered air-purifying respirators may use a loose-fitting hood, hat, or facepiece that forms only a partial seal with the face, a good option for an individual with facial deformities or facial hair (Fig. 1- 7). These loose garments also may incorporate other protective features. The headpiece may incorporate a helmet. A hood may consist of material that when draped over the upper body provides additional protection from blasting grit or chemical hazards.

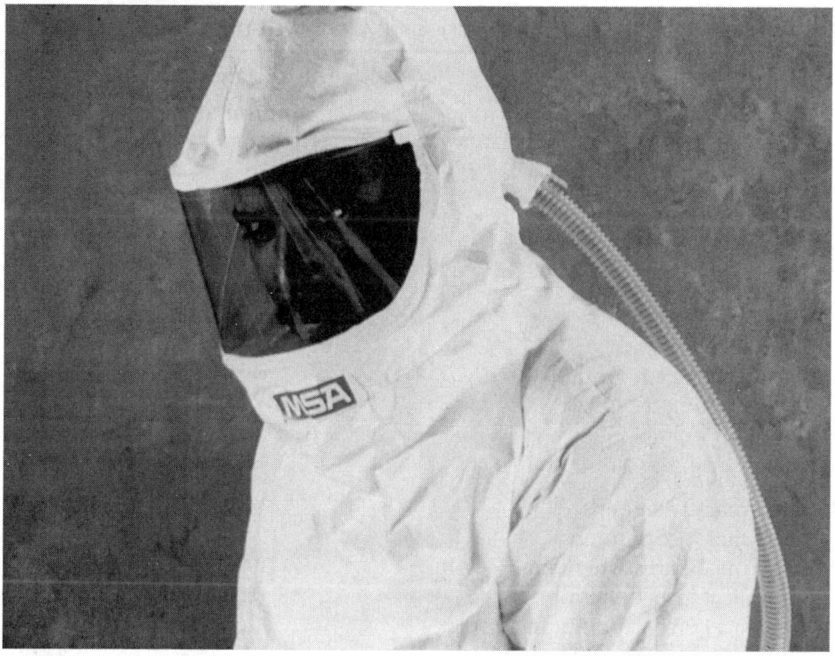

Figure 1-7. Supplied-air loose-fitting respirator. (Reproduced courtesy of MSA.)

Protection Factors

Both the American National Standards Institute (ANSI) and NIOSH rate respirators by their ability to reduce a worker's exposure to the ambient environment's concentration of contaminants. As with other personal protective equipment, the actual protectiveness of a respirator in use varies from its theoretical capabilities. *Assigned protection factor* (APF) refers to the potential protectiveness of a respirator based on its design and laboratory performance. An APF of 10 means that the device reduces by tenfold the worker's exposure to the ambient concentration of contaminants. APF ratings range from 10 for a negative-pressure air-purifying respirator up to 10,000 for a self-contained breathing apparatus (SCBA) operating in pressure-demand mode (Table 1-3). APF ratings have not yet been established for particulate-filtering respirators. NIOSH recommends that employers use APF ratings when choosing the appropriate respirator for a specific environment. Nonetheless, NIOSH warns that actual protection in the field may be considerably less. *Measured protection factor* (MPF) describes a respirator's performance quantified under optimal conditions in the field. *Workplace protection factor* (WPF) expresses actual protection achieved under usual work

Table 1-3. Table of APFs for Various Types of Respirators

Respirator class and type	OSHA cadmium Std.	NIOSH
Air-purifying		
Filtering facepiece	10	10
Half-mask	10	10
Full facepiece	50	50
Powered air-purifying		
Half mask	50	50
Full facepiece	250	50
Loose-fitting facepiece	25	25
Hood or helmet	25	25
Supplied air		
Half mask, demand	10	10
Half mask, continuous	50	50
Half mask, pressure demand	1000	1000
Full facepiece, demand	50	50
Full facepiece, continuous flow	250	50
Full facepiece, pressure demand	1000	2000
Loose-fitting facepiece	25	25
Hood or helmet	25	25
Self-contained breathing apparatus (SCBA)		
Demand	50	50
Pressure demand	>1000	10,000

Source: Occupational Safety and Health Administration. Assigned Protection Factors. Reprinted from a table available at http://www.osha-sic.gov/SLTC/respiratory_advisor/haz_expose/apf/apf.html

conditions. As additional research creates MPF and WPF data, ANSI and NIOSH update their published APF ratings. OSHA is developing a new regulation to address APF ratings. In the interim, OSHA recognizes the APF ratings declared in substance-specific standards, specific OSHA interpretations, or the NIOSH APF ratings.

Psychophysiologic Demands of Respirators

The wide range of respiratory protection devices interacts with multiple conditions of use to impose complex and varying psychophysiologic stresses on workers. Respirator components that contribute to these stresses consist of the mask, air-purification medium, inflow and outflow valves, and method of air supply. These components contribute their burden by adding respiratory dead space, flow resistance, weight, thermal insulation, and facial enclosure. In turn, these burdens create stressors, such as carbon dioxide retention, enhanced cardiorespiratory work, musculoskeletal strain, sensory impairment, increased thermal load, and interference with drinking or eating that can lead to symptoms.[5] To be effective, respirators must be tolerated by their users and allow them to accomplish their job tasks despite these encumbrances. No simple chart can be drawn that defines the actual demands placed on a worker in the field. Nonetheless, a general scheme can be outlined that helps explain the impact that different respirator components may have on an individual (Table 1-4).

Respirators limit sensory input and communication. The physical bulk of a respirator decreases the visual field. Dirty and fogged face shields may impair visual acuity. Respirator noise and hoods decrease hearing. All respirators cover the mouth and thereby decrease voice clarity and volume.[6] They minimize olfactory cues about environmental events. These effects on vocal expression and sensory input interfere with the ability to identify safety hazards and alert others to their existence. They also can impede job functioning. Quarter- and half-mask negative-pressure respirators with integral purifying elements minimize sensory and communication stresses.

The physical contact of a respirator on the face can lead to allergic or irritant responses of the skin.[7] Increases in facial skin temperature, along with the accumulation of sweat, inside a full-facepiece respirator mask degrade its acceptability.[8] Physical forces around the orbits may result in difficulties wearing contact lenses. Moreover, the need for a tight seal limits the types of corrective lenses that can be used. In addition, such masks are incompatible with facial hair and some facial deformities. Loose-fitting respirators minimize the physical effects associated with devices requiring a tight seal.

A respirator may be uncomfortable not only because of the physical pressure on the face but also because it interferes with a worker's habits such as smoking, chewing gum, or scratching. Its presence impedes the concurrent administration of medications, maintaining hydration, or eating. Further, workers cannot easily and safely attend to such problems as rhinorrhea, cough, and eye irritation while wearing a respirator. The tendency for respirators to shift habitual breathing patterns from predominantly nasal to oral flow also probably contributes substantially to respirator intolerance.[9]

Table 1-4. Relationships Between Respirator Components, Physiologic Stress, and Symptoms

Component	Physiologic load	Physiologic stress	Symptom
Mask/face piece	Dead space	CO_2 retention	Dyspnea
		Cardiovascular work	Angina/arrhythmia/CHF[a]
	Facial enclosure	Impaired special senses	Decreased environmental awareness and ability to communicate
		Inability to eat/drink/spit	Hunger, thirst, hygiene
			Claustrophobia
		Dermal contact	Facial irritation, latex allergic symptoms
	Thermal insulation	Heat stress	Fatigue/mental clouding
Airflow valves and air-purifying devices	Flow resistance	CO_2 retention	Dyspnea
		Cardiovascular work	Angina/arrhythmias/CHF
		Pneumothorax	Pleuritic chest pain
			Dyspnea
Air supply	Weight	Cardiovascular work	Angina/Arrhythmia/CHF
		Heat stress	Fatigue/mental clouding
		Impaired mobility	Musculoskeletal distress
	Thermal insulation	Heat stress	Fatigue/mental clouding
	Air pressure bias	Cardiovascular work	Angina/arrhythmia/CHF
		CO_2 retention	Dyspnea

[a]CHF = congestive heart failure.
Source: Adapted from Muhm JM. medical surveillance for respirator users. *J Occup Environ Med* 1999;41:992.

Respirators add to thermal stress by their weight, head enclosure, and method of air supply. In particular, closed-circuit SCBAs, which produce oxygen by an exothermic reaction, and full-face negative-pressure respirators are culprits. The impervious protective clothing frequently worn with these respirators aggravates the thermal burden.[10,11] Although the cooling effect of PAPRs can be welcome in hot environments, the opposite is true in cold environments.

Respiratory protection devices augment the work of ventilation by increasing breathing resistance and to a lesser extent by enlarging respiratory dead space.[12,13] As a result, workers can experience alterations in breathing patterns, hypoventilation, retention of carbon dioxide, and a decrease in their exercise capacity.[14] Air-purifying negative-pressure and atmosphere-supplying demand-type respirators increase inspiratory resistance. All respirators provide some expiratory resistance because the wearer must work to open the exhalation valve. Pressure-demand respirators and SCBAs have exhalation valves designed to maintain positive pressure within the mask. These respirators provide the greatest resistance to exhalation. The functional dead space volume of a respirator consists of the volume of air within the respirator that mixes with inhaled and exhaled air. This respiratory dead space increases the amount of air that an individual must breathe (the minute ventilation) to maintain arterial carbon dioxide tension.

The airway resistance associated with negative-pressure respirator use limits the maximum short-term and 8-hour exercise capacity and endurance of healthy men.[15,16] At submaximal workloads, however, research suggests that no clinically significant impairment of normal respiratory function occurs in healthy or even mild to moderately pulmonary-impaired subjects using properly maintained negative-pressure respirators.[17–19] As workloads increase, workers with respiratory impairment can be expected to be more susceptible to the additional physiologic stresses of a respirator.[19] In addition, field conditions and poor respirator maintenance exacerbate adverse respiratory effects measured in the laboratory.[4]

Theoretically, the greater fluctuations in intrathoracic pressure that accompany the increased work of breathing associated with respirators may increase the risk of pneumothorax. Actual pleural pressures, however, remain less than that of a vigorous cough.[20]

Respirators increase the load on the cardiovascular system. Increased work of breathing contributes minimally to the cardiac work compared with the physical burden and constraints of some types of respirators.[21–24] Nonetheless, statistically significant blood pressure, heart rate, and respiratory rate changes occur even in healthy individuals wearing disposable low-resistance negative-pressure respirators during heavy work after periods as short as 5 minutes.[13] Some of these changes are noted at light and moderate workloads as well. SCBAs, which may weigh up to 40 lb, decrease maximum work capacity by as much as 20 percent and increase heart rate and oxygen consumption at submaximal exertion.[25]

Bulky, heavy respirators add to the musculoskeletal stresses of work, increasing the likelihood of injury. Hoses, masks, and associated equipment decrease mobility and add to the risk of trips, falls, and possible entanglement. One study found that full-face respirators routinely impair equilibrium at any workload after use for as few as 15 minutes.[26]

In the end, respirator tolerance must be measured by whether an individual can perform a job while wearing a respirator. Evidence is mixed as to whether cognitive faculties are affected.[27,28] Workers using respirators cannot perform at the same physiologic level as those without respiratory devices. Physiology alone, however, cannot adequately explain respirator tolerance. Indeed, other factors appear to predominate, especially at lower levels of exertion.[4,29] The interplay of various psychophysical sensations created by a respirator overwhelms some workers' capacity to tolerate the device.[30–32] Even beliefs about discomfort and inconvenience were found to be the major determinant of respirator compliance in one study.[33] Morgan[34] estimates that 10 percent of "normal" young men experience dizziness, claustrophobia, or anxiety attacks while exercising during respirator use. Anxious subjects wearing respirators experienced more discomfort, perform for shorter times, and accomplish less total work than their lower-anxiety counterparts.[35]

Because respirators reduce endurance and maximal work capacity, are uncomfortable, and simply get in the way, they affect work performance and the risk of other injuries.[36] One study found that high-efficiency particulate air (HEPA) respirators affected communication, visibility, and range of motion, resulting in a significant increase in the number of contaminated sharps injuries among health care workers.[37] Medical examiners must recognize that respirator noncompliance resulting from these factors looms larger than the risk of specific adverse health effects such as a myocardial infarction or exacerbation of an asthmatic attack. Removal of a respirator in a hazardous environment will result in exposure that can cause illness or death.

Job Demands

Respirators place psychophysiologic demands on workers that become increasingly significant with heavier and bulkier respirators. These burdens become greater with the addition of other personal protective equipment, a warm environment, and the exertion required by the job tasks.[10,11] At the same time, objective protection factors of respiratory protective devices are reduced under warmer conditions.[38] It makes no sense to approve someone for respirator use without understanding the nature of the job to be accomplished while wearing a respirator. Will the person sit in an air-conditioned place pushing buttons, fight an interior structural fire, or act as a first responder to a hazardous waste spill? People wearing respirators in environments that are immediately dangerous to life or health (IDLH) have no margin for respirator failure. Further, an individual's personal health condition may eliminate a respirator as an appropriate approach to protection from a specific hazard. For example, a respirator adequately protective for most people for an airborne hazard may not protect an individual sensitized to the agent.[12] A person sensitized to toluene diisocyanate may have no problems tolerating the respirator but may not be adequately protected by the respirator if asked to work with the substance. Work variables important to the examining clinician include other personal protective equipment, aerobic demands, thermal environment, chemical hazards and their airborne concentration, frequency and duration of respirator use, and work at unprotected heights.

Conclusion

Although respiratory protective devices should take last place in a program to protect workers from respiratory hazards, millions of workers use them on a regular basis. They can only be protective in the context of a comprehensive respiratory protection program. A case of respirator failure indicates the need to review an entire program. As part of this program, medical evaluation helps assess an employee's susceptibility to the psychophysiologic burden associated with respirator use. Job tasks and environmental conditions typically impose a more significant psychophysiologic load on the respirator user; factors to be considered include the type of respirator, the duration and frequency of its use, the level of physical work effort, the use of protective clothing, and the presence of temperature extremes or high humidity. The quality of the medical evaluation depends on the examiner's understanding of the specific demands placed on an individual by different respirators and the conditions under which they will be used.

Chapter 2 walks through the steps an employer should take before referring an employee for a medical evaluation for respirator use.

Further Reading

Medical Surveillance

Harber P. Surveillance and screening of respiratory disease. In: Harber P, Schenker MB, and Balmes JR (eds), Occupational And Environmental Respiratory Disease. St. Louis: Mosby, 1996:837–855. This chapter provides a good overview of the purpose and methods of respiratory medical surveillance.

Respiratory Protection Devices

Bollinger NJ, Schutz RH. *NIOSH Guide to Industrial Respiratory Protection,* DHHS (NIOSH) Publication No. 87-116. Cincinnati: U.S. Department of Health and Human Services, Public Health Service, Centers for Disease Control, National Institute for Occupational Safety and Health, Division of Safety Research, 1987. *This monograph provides a comprehensive yet easy-to-understand overview of respirators, their selection, and their use.*

Colton CE, Birkner LR, Brosseau LM. *Respiratory Protection: A Manual and Guideline,* 2d ed. Akron: American Industrial Hygiene Association, 1991. *This manual offers a complete, detailed explanation of respiratory protection programs and recommendations for instituting their various components.*

Harber P. Respirators. In: Rom WN (ed), *Environmental and Occupational Medicine,* 3d ed. Philadelphia: Lippincott-Raven, 1998:1757–1765. *Harber summarizes the science of respirators and respiratory protection programs.*

Muhm JM. Medical surveillance for respirator users. *J Occup Environ Med* 1999; 41:989–994. *Muhm simplifies the complexities of respirators and offers a logical approach to understanding their impact on human health, with excellent suggestions for providing accommodations to workers who require them.*

Revoir WH, Bien C. *Respiratory Protection Handbook.* Boca Raton, FL: CRC Press, 1997. *This book provides a highly technical and comprehensive overview of the science of respiratory protective devices.*

OSHA Rationale for Medical Evaluation

OSHA Preambles. Respiratory Protection: VII. Summary and Explanation. Available at *http://www.osha-slc.gov/Preamble/RP_html/RESPIRATORY7.html. This OSHA preamble reviews the agency's scientific logic for the revised Respiratory Protection Standard.*

References

1. Respiratory Protection: VI. Summary of the Final Economic Analysis. Available at *http://www.osha-slc.gov/Preamble/RP_html/RESPIRATORY6.html#TEXT1.*
2. Conway H, Simmons J, Talbert T. The purposes of occupational medical surveillance in US industry and related health findings. *J Occup Med* 1993;35:670–686.
3. Revoir WH, Bien C. *Respiratory Protection Handbook.* Boca Raton, FL: CRC Press, 1997:357–358.
4. Harber P. Respirators. In: Rom WN (ed), *Environmental and Occupational Medicine,* 3d ed. Philadelphia: Lippincott-Raven, 1998:1757–1765.
5. Muhm JM. Medical surveillance for respirator users. *J Occup Environ Med* 1999;41:989–994.
6. Coyne KM. Respirators performance ratings for speech intelligibility. *Am Ind Hyg Assoc J* 1998; 59:257–260.
7. Elmer KB. Contact urticaria to the MCU-2A/P gas mask. *Milit Med* 1999; 164:377–378.
8. Johnson AT. Sweat rate inside a full-facepiece respirator. *Am Ind Hyg Assoc J* 1997; 58:881–884.
9. Harber P. Study of respirator effect on nasal-oral flow partition. *Am J Ind Med* 1997; 32:408–412.
10. White MK, Vercruyssen M, Hodous TK. Work tolerance and subjective responses to wearing protective clothing and respirators during physical work. *Ergonomics* 1989; 32:1111–1123.
11. White MK, Hodous TK, Vercruyssen M. Effects of thermal environment and chemical protective clothing on work tolerance, physiological responses, and subjective ratings. *Ergonomics* 1991; 34:445–457.
12. Beckett WS. Certifying the worker for respirator use. *Semin Occup Med* 1986; 1:119–124.
13. Jones JG. The physiological cost of wearing a disposable respirator. *Am Ind Hyg Assoc J* 1991; 52:219–225.
14. Louhevaara VA. Physiological effects associated with the use of respiratory protective devices: A review. *Scand J Work Environ Health* 1984; 10:275–281.

15. Deno NS, Kamon E, Kiser DM. Physiological responses to resistance breathing during short and prolonged exercise. *Am Ind Hyg Assoc J* 1981; 42:616–622.
16. Sulotto F, Romano C, Dori S, et al. The prediction of recommended energy expenditure for an 8 h work day using an air-purifying respirator. *Ergonomics* 1993; 36:1479–1487.
17. Harber D, Tamimie J, Bhattacnarya A, Barber M. Physiologic effect of respirator dead space and resistance loading. *J Occup Med* 1982; 24:681–684.
18. Hodous T, Petsonk L, Boyles C, et al. Effects of added resistance to breathing during exercise in obstructive lung disease. *Am Rev Respir Dis* 19XX; 128:943–948.
19. Raven PB, Jackson Aw, Page K, et al. The physiological responses of mild pulmonary impaired subjects while using a "demand" respirator during rest and work. *Am Ind Hyg Assoc J* 1981; 42:247–257.
20. *NIOSH Respirator Decision Logic,* DHHS (NIOSH) Publication No. 87-108. Cincinnati: US Department of Health and Human Services, Public Health Service, Centers for Disease Control, National Institute for Occupational Safety and Health, Division of Safety Research, 1987:53.
21. Louhevaara V, Smolander J, Tuomi T, et al. Effects of an SCBA on breathing pattern, gas exchange, and heart rate during exercise. *J Occup Med* 1985; 27:213–216.
22. Louhevaara V, Tuomi T, Korhonen O, Jaakkola J. Cardiorespiratory effects of respiratory protective devices during exercise in well-trained men. *Eur J Appl Physiol* 1984; 52:340–345.
23. Gee JBL, Burton G, Vassallo C, Gregg J. Effects of external airway obstruction on work capacity and pulmonary gas exchange. *Am Rev Respir Dis* 1968; 98:1003–1012.
24. Verstappen F, Bloemen L, Van Putten M, Reuvers J. Self-contained respirators: Effects of negative and positive pressure-demand types on physical exercise. *Am Ind Hyg Assoc J* 1986; 47:635–640.
25. Raven PB, Davis TO, Shafer CL, Linnebur AC. Maximal stress test performance while wearing a self-contained breathing apparatus. *J Occup Med* 1977; 19:802–806.
26. Seliga R, Bhattacharya A, Succop P, et al. Effect of work load and respirator wear on postural stability, heart rate and perceived exertion. *Am Ind Hyg Assoc J* 1991; 52:417–422.
27. Caretti DM. Cognitive performance during long-term respirator where while at rest. *Am Ind Hyg Assoc J* 1997; 58:105–109.
28. Caretti DM. Cognitive performance and mood during respirator wear and exercise. *Am Ind Hyg Assoc J* 1999; 60:213–218.

29. Harber P, Beck J, Brown C, Luo J. Physiologic and subjective effects of respirator mask type. *Am Ind Hyg Assoc J* 1991; 52:357–362.
30. Shimozaki S, Harber P, Barrett T, Lossides P. Subjective tolerance of respirator loads and its relationship to physiological effects. *Am J Ind Hyg Assoc J* 1988; 49:108–116.
31. Harber P, Shimozaki S, Barrett T, Loisides P. Relationship of subjective tolerance of respirator loads to physiologic effects and psychophysical load sensitivity. *J Occup Med* 1989; 31:681–686.
32. Chia SE. A study on the usage of respirators among granite quarry workers in Singapore. *Singapore Med J* 1989; 30:269–272.
33. White MC, Baker EL, Larson MB, Wolford R. The role of personal beliefs and social influences as determinants of respirator use among construction painters. *Scand J Work Environ Health* 1988; 14:239–245.
34. Morgan WP. Psychological problems associated with a wearing of industrial respirators: A review. *Am Ind Hyg Assoc J* 1983; 44:671–677.
35. Johnson AT. Influence of anxiety level on work performance with and without a respirator mask. *Am Ind Hyg Assoc J* 1995; 56:858–865.
36. Johnson AT. Effect of altered vision on constant load exercise performance while wearing a respirator. *Am Ind Hyg Assoc J* 1997; 58:578–582.
37. Eck EK. The effect of high-efficiency particulate air respirator design on occupational health: A pilot study balancing risks in the real world. *Infect Control Hosp Epidemiol* 1997; 18:122–127.
38. Meyer JP. Field study of subjective assessment of negative pressure half-masks: Influence of the work conditions on comfort and efficiency. *Appl Ergon* 1997; 28:331–338.

2

Before the Medical Evaluation: An Overview of the Employer's Responsibilities

The medical evaluation plays only one part in a much larger endeavor to protect worker health from airborne hazards. At a minimum, the Occupational Safety and Health Administration (OSHA) mandates employer compliance with its revised respiratory protection standard. The respiratory protection standard (29 CFR 1910.134), originally promulgated in 1971, aims to protect workers from occupational respiratory hazards. By themselves, respirators frequently fail in their goal of preventing exposure to airborne hazards. To prevent disease, employers must use respirators, like other personal protective equipment (PPE), in the context of a comprehensive respiratory protection program. At best, respirators provide personal protection from respiratory hazards. As with other PPE, OSHA requires that respiratory protection devices remain at the end of a hierarchy of industrial hygiene controls to reduce workers' exposures. Whenever feasible, the standard mandates the preferential use of less hazardous materials, engineering controls, and work-practice modifications. This chapter reviews employer responsibilities for respiratory protection that are most relevant to the physician and licensed health care professional (PLHCP).

Comprehensive respiratory protection programs that include engineering and administrative controls undeniably have decreased occupational diseases such as asbestosis, silicosis, and aluminum worker pulmonary diseases. Research on the contributions of respirators to these successes generally has relied on indirect measures of effectiveness: (1) whether respirators are used properly and (2) whether they reduce exposure as measured by biologic monitoring.[1]

Respirators may fail to protect workers for a variety of reasons (Table 2-1). Although equipment failure and psychophysiologic intolerance can occur, human error and noncompliance more commonly cause

Table 2-1. Reasons for Respirator Failure

Inadequate environmental engineering
Incorrect selection of respirator
Psychophysiologic intolerance associated with
Incomplete medical evaluation
Inaccurate medical recommendation
Failure to reevaluate worker if health status or work conditions change
Poor respirator fit
Inadequate training
Improper cleaning and maintenance
Equipment failure
Noncompliance

problems. Programmatic features, including incorrect respirator choice, poor fit, inadequate maintenance, nonavailability of respirator when needed, unclean storage, and insufficient training, may be at fault. Respirators can only succeed as part of a complete respiratory protection program.

Unfortunately, several surveys suggest that program deficiencies are common.[1] A study of a respiratory protection program for lead workers showed that they often used incorrect respirators or no respirators at all.[2] In 1997, a telephone survey concluded that respirator maintenance programs usually were not adequate.[3] OSHA's inspection data reveal similar problems. An analysis of OSHA inspection data from 1976 through 1982 found that when respirators were used to protect employees from excessive levels of air contaminants, respiratory protection programs usually were deficient in one or more elements.[4] Violation of the OSHA respiratory protection standard was the fifth most frequent citation by OSHA officials in the period from October 1998 through September 1999.[5] OSHA has concluded that employer noncompliance with the older respiratory protection standard was a critical factor in at least 47 deaths and 126 catastrophic injuries in 1994–1995.[4] The most frequently cited deficiencies included failure to provide respirators at all, absence of standard operating procedures governing respirator use, and neglect to train or fit test respirator users adequately. Compliance data from 1990–1996 document that over 72 percent of companies inspected for an overexposure incident did not comply with the respiratory protection standard.[4]

The combination of weaknesses in the original respiratory protection standard, poor compliance, and the evolution of respirator technology led OSHA to the first major revision of the respiratory standard in 25 years. The new respiratory protection standard promulgates the use of more effective respirators, provides guidelines for respirator selection and use, clarifies responsibilities, and provides employers with direction regarding hazard assessment, medical evaluation, fit testing, and employee training.[6] Compliance with all aspects of the revised 29 CFR 1910.134 helps meet the goal of preventing occupational disease (Table 2-2).

Who Is Covered?

The standard applies to General Industry (Part 1910), Shipyards (Part 1915), Marine Terminals (Part 1917), Longshoring (Part 1918), and Construction (Part 1926). Previously scattered throughout OSHA health standards, all respirator-related provisions have been consolidated in CFR 1910.134. The Environmental Protection Agency's (EPA's) worker protection standard (40 USC Part 170) covers respiratory protection in the agricultural industry. Presently, employers with workers potentially exposed to tuberculosis need only comply with the original respirator standard, redesignated 29 CFR 1910.139.

The OSHA standard does not preempt other relevant federal agency standards such as the Nuclear Regulatory Commission (NRC) or the Department of Transportation's (DOT's) Office of Pipeline Safety. In an interpretive letter,[7] OSHA explains

> Section 4(b)(1) of the Occupational Safety and Health Act, 29 U.S.C. § 653(b)(1), precludes OSHA from applying its standards to working conditions that are regulated by other federal agencies. In order for a working condition to qualify for the exemption, the other federal agency must have statutory authority to regulate the health and safety of working conditions of employees and must exercise that authority by standards or regulations having the force and effect of law. Section 4(b)(1) does not create an industry-wide exemption. It only exempts specific "working conditions" that are subject to the worker safety or health regulations of other agencies.

The conditions of this interpretation do not relieve any employer of the requirement to ensure that its employees receive a medical

Figure 2-2. 29 CFR Part 1910.134: Elements of a Respiratory Protection Program

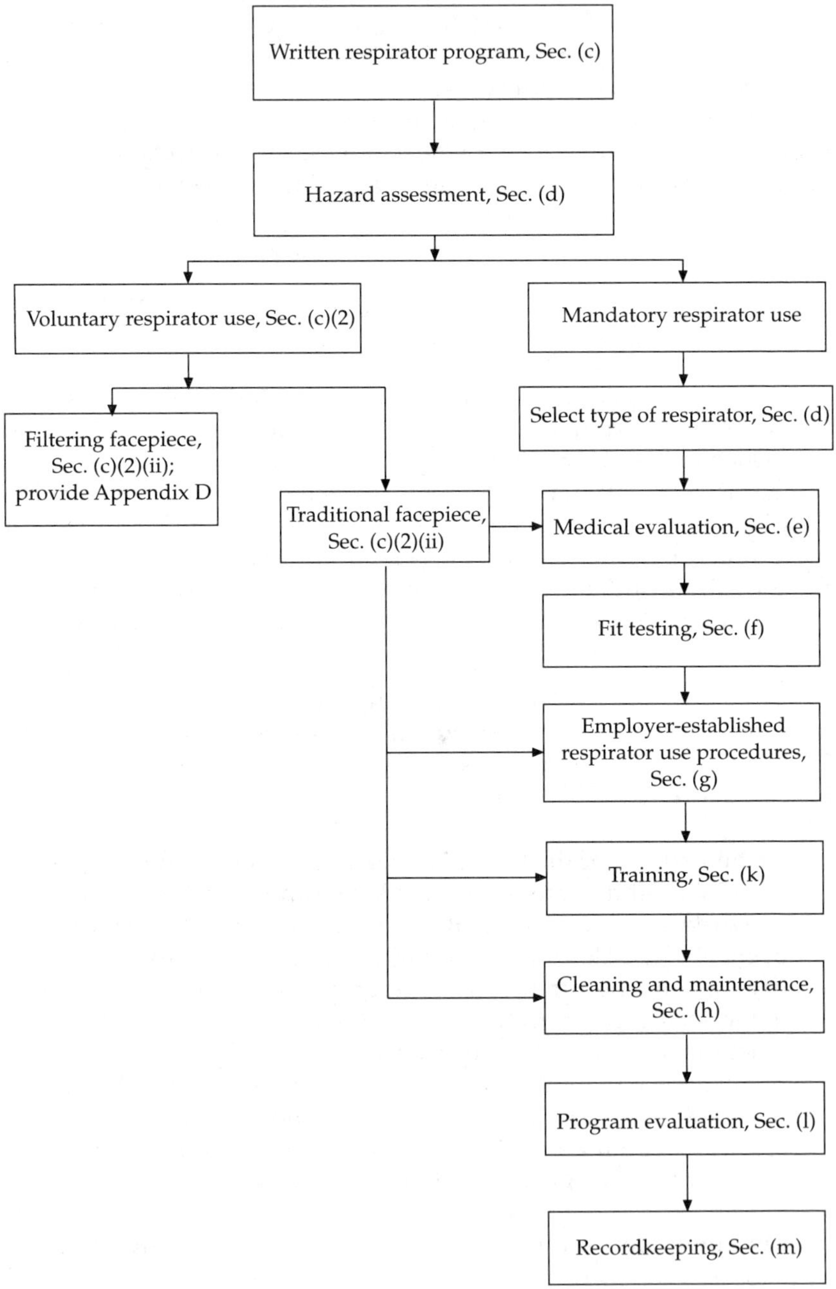

evaluation prior to respirator use. Some agencies, however, may place additional requirements on medical evaluation. Physicians and LHCPs are urged to refer to agency-specific regulations for more information.

The revised standard requires that employers provide respirators, training, and medical evaluation to their workers whenever necessary to protect their health. In this instance, a company must maintain a comprehensive, written respiratory protection program implementing all the standard's elements. The respirator requirement may stem from either OSHA regulation or internal company policy.

Employers also may provide respirators at the request of employees or permit employees to wear their own respirators in settings where no respirator requirement exists. Defined as "voluntary use" by OSHA, this circumstance requires that the company first determine that such respirator use in itself will not create a hazard. The employer also must provide voluntary respirator users with "Information for Employees Using Respirators When Not Required under the Standard," found in Appendix D of the standard (Appendix 1D of this book). Otherwise, the company must only implement those aspects of the written respiratory protection program necessary to ensure that voluntary use does not present a health hazard to the user. Aspects include a medical evaluation prior to use and appropriate respirator cleaning, storage, and maintenance. Voluntary use of filtering facepieces (dust masks) does not require a written respiratory protection program.[8]

Respiratory Protection Program Administrator

29 CFR 1910.134 requires employers to designate a program administrator who is "qualified by appropriate training or experience that is commensurate with the complexity of the program. . . ." The program administrator oversees compliance with all aspects of the respiratory protection program: hazard assessment, respirator selection, respirator care and maintenance, medical evaluation, fit testing, employee training, respirator use, program assessment, and record keeping.[9] During an inspection, an OSHA compliance officer may examine the qualifications of the program administrator, the administrator's knowledge of the respiratory protection standard, the company's written respiratory protection program, and the appropriate use of respirators at the work site.[10]

Respiratory Protection Program

The complexity of a company's respiratory protection program depends on the number, type, and concentrations of air contaminants; other environmental conditions; the job demands; and the concurrent use of other PPE. A company's written program must include each of the following nine elements:

1. Procedures for selecting respirators
2. Medical evaluation of workers required to use respirators
3. Fit-testing procedures for tight-fitting respirators
4. Respirator use procedures for both routine and foreseeable emergency activities
5. Respirator maintenance procedures and schedules (cleaning, disinfecting, storage, repair, and discarding)
6. Procedures for ensuring the adequacy of air quality, quantity, and flow for atmosphere-supplying respirators
7. Training for employees regarding the respiratory hazards in the workplace
8. Training for employees on the limitations of the assigned respirator and its proper fit, use, and maintenance
9. Procedures for evaluating the effectiveness of the company's respiratory protection program

The following sections briefly address each of the nine elements of a written respiratory protection program. For more information about compliance with the OSHA standard and respiratory protection in general, the reader is referred to the additional readings listed at the end of this chapter.

Procedures for Selecting Respirators

OSHA obligates the employer to select respirators specific to the respiratory hazard(s) in the workplace and associated environmental and user factors. Hazard assessment therefore must precede respirator selection. OSHA specifies the general types of respirators acceptable for use in different atmospheres and work settings, such as an environment that is immediately dangerous to life and health (IDLH). All respirators selected must be certified by the National Institute of

Occupational Safety and Health (NIOSH) and used in compliance with the conditions of NIOSH certification. The employer must offer the employee a sufficient number of models and sizes of the hazard-appropriate respiratory protection device to optimize comfort and fit.

Hazard Assessment

Prior to respirator selection, the employer must conduct a comprehensive review of all worker exposures, job demands, and potential adverse effects of their interactions. Hazard assessment begins with an analysis of work-site air contaminants and the potential for their control through engineering and administrative methods. This initial step provides the employer with data to decide whether respirators are necessary to protect worker health. Employers always must first institute feasible engineering and administrative controls to minimize potential worker exposures.[10] OSHA condones respirator use only (1) when engineering and work-practice controls are not sufficient to reduce exposure to a permissible level, (2) when the institution of controls is pending, and (3) in emergency situations.[10]

The assessment includes both air monitoring and identification of contaminant characteristics: physical, chemical, and toxicologic attributes; warning properties; exposure limits; the IDLH concentration; eye irritancy; and the hazard it presents to the worker under normal, temporary (e.g., maintenance), and emergency conditions. Sources for obtaining exposure-limit data include NIOSH's recommended exposure limit (REL), OSHA's permissible exposure limit (PEL), the American Congress of Governmental Industrial Hygienists' threshold limit values (TLVs), and other human exposure limits established by qualified professionals.

Under certain circumstances, OSHA permits some flexibility in worker exposure assessment.[4] Acceptable approaches include monitoring personnel, monitoring work-site areas, using exposure data from laboratory or similar industry studies, and predicting from exposure-modeling equations. When using estimates, OSHA expects the employer to take into account exposure variability. Estimates should derive from studies that either account for variability or use worse-case assumptions.[4] If laboratory or other industry data are used, the employer must review the study conditions for appropriate workplace similarities such as the industrial process, types of materials, control methods, and environmental conditions. Modeling equations may be used for assessing respiratory exposures when workplace factors remain reasonably constant and predictable.[10] OSHA also permits exposure estimates for such activities as construction, where exposure is short term and "monitoring of specific employees or operations is not available when respirators are selected." In this instance, OSHA requires reasonable and conservative exposure estimates based on past monitoring or other representative data.[10] When exposure-monitoring data or reasonable estimates are not available, employers must regard the work environment as immediately dangerous to life and health (IDLH).

Beyond the analysis of airborne contaminants, the hazard assessment must include a review of other occupational hazards, the general environmental conditions, and the physical demands of the job task. Important data include the amount of energy expended for task completion (Table 2-3), as well as the frequency and duration of the job task. Ambient factors such as heat, cold, and humidity affect not only worker comfort and health but also the service life of respirator cartridges and canisters. An environment with loud noise exposures can further impair communication already hampered by respirators. Other considerations include the concurrent use of other PPE, awkward positions, mobility, interfering work equipment, and user comfort. Safety factors, such as confined spaces, heights, and potentially

Table 2-3. Energy Requirement for Various Activities

METs level	Activity
1–2 METs	Driving, desk work, standing, slow-paced walking
2–3 METs	Custodial work, mechanic
3–4 METs	Window washing, machine assembly, welding, brick laying
4–5 METs	Painting, masonry, light carpentry
5–6 METs	Light shoveling, spading garden
6–7 METs	Shoveling 10 lb
7–8 METs	Carrying 80 lb, heavy carpentry, digging ditches
8–9 METs	Shoveling 14 lb, heavy machinery assembly
10+ METs	Shoveling 16 lb, firefighting

Source: Adapted from Temte J. Cardiovascular conditions and worker fitness and risk. *Occup Med* 1988; 3:244.

interfering work equipment, also play a role in respirator selection. Air-line respirators, for example, present intolerable risks and inconvenience for jobs that require a high degree of mobility or when other work equipment may catch or tangle the hose.[11]

Hazard assessment is an ongoing process that helps keep a company's respiratory protection program relevant to contemporary conditions at the work site. Changes in the work environment, workload, or work process, as well as changes in the type of air contaminants, may require revision of a company's respiratory protection plan.[10]

Respirator Selection

CFR 1910.134 prescribes general parameters for the types of respirators required in certain environments. *NIOSH Respirator Decision Logic* provides the rationale and algorithm for identifying the appropriate respirator for use in a particular environment.[12] Since 42 CFR Part 84 creates nine new classes of particulate filters, the original NIOSH decision logic must be supplemented with an algorithm for selecting the correct particulate filter. The Industrial Safety Equipment Association (ISEA) offers a decision-making flowchart that considers the nature of the particulate contamination in the workplace.[13] Correct respirator selection rests on the validity of the hazard assessment. The NIOSH decision tree relies on

- General use conditions
- Concentrations of contaminants
- The physical, chemical, and toxicologic aspects of the air contaminants
- The contaminants warning properties
- Exposure limits
- IDLH concentrations
- Eye irritation potential
- Limitations of each class of respirator

The NIOSH selection process considers respirator characteristics such as service life, information about cartridges and canisters, purification efficiency, air supply capability, face seal characteristics, and assigned protection factors (APFs) for each class of respirator. Organizations such as the American National Standards Institute

(ANSI), NIOSH, and OSHA have assigned somewhat different protection factors. OSHA plans to publish its own APFs as part of the respiratory protection standard. Meanwhile, employers should rely on the APFs listed in OSHA's substance-specific standards, if applicable, or APFs granted by a specific OSHA interpretation or the NIOSH APFs[14] (see Table 1-3). Employers should choose respirators with the realization that actual protection (workplace protection factor, or WPF) falls below the published APF (see Chapter 1).

Medical Evaluation

OSHA has concluded that using a respirator places a psychophysiologic burden on employees that varies with the type of respirator worn, the job and workplace conditions in which the respirator is used, and the medical status of the employee. Accordingly, the agency mandates a medical evaluation to assess an employee's ability to use a respirator before the employee is fit tested or required to use the respirator in the workplace. Employers must provide a medical evaluation, at no cost to the employee, for every worker required to use a respirator, regardless of the duration or frequency of respirator use. During an audit, OSHA may interview physicians and LHCPs, as well as individuals at the work site, regarding the medical evaluation and confidentiality of the process.[10]

OSHA has published interpretive letters regarding its requirement for medical evaluation. The agency does not require that employers provide a medical evaluation for employees who use a respirator for escape purposes only.[15] Employees need not submit to a medical evaluation, but they cannot be assigned to an area where a respirator is required without a medical evaluation.[16] The employer also must provide a medical evaluation for employees who voluntarily wear elastomeric facepieces for comfort or other reasons where no hazard exists.[17] Only employees who voluntarily wear filtering facepieces (dust masks or disposable dust respirators) need not be evaluated medically.[18] Under the Hazard Communication Standard [29 CFR 1910.1200(c)], employers must advise contractors and visitors of hazards to which they may be exposed, but businesses do not have a responsibility to medically evaluate such individuals to wear respirators. This responsibility falls back on the employers of the contractors and visitors.[19]

29 CFR 1910.134(e) specifies the minimum requirements for the medical evaluation (Appendix 1). The revised standard helps clarify the medical evaluation process and communication requirements for

both employers and health care personnel. The most significant changes consist of options about how to conduct the evaluation and who can perform it. The employer may elect to use the medical questionnaire in Appendix C of the standard (Appendix 1C of this book) or a traditional medical examination that contains at least the elements included in the questionnaire. Chapter 3 discusses the administrative and clinical details of the medical evaluation.

Choosing a Physician or Licensed Health Care Provider

The standard places the burden of identifying an appropriate health care provider on the employer. Successful and effective implementation of the medical evaluation process requires an understanding of several key issues on the part of both the employer and the health care provider. Foremost, the health care provider must come to the job with an understanding of respirators and their use, as well as their physiologic, psychological, and safety impacts on workers.[9] Employers should interview health care providers about their knowledge and decision-making logic in qualifying workers for respirator use.

The revised standard permits either a physician or a licensed health care provider (LHCP) to conduct the medical evaluation. LHCPs can only work within the scope of practice defined by their state license. Every state defines the credentials, responsibilities, level of autonomy, and limitations for each health care professional's discipline.[8] Whether LHCPs can conduct the entire medical evaluation or only a portion depends on their scope of practice as defined by the state in which they work.

On August 4, 1999, NIOSH raised policy concerns about "unsupervised medical evaluations conducted by health care professionals who are not licensed for independent practice to perform or supervise medical evaluations."[20] The American College of Occupational and Environmental Medicine unsuccessfully petitioned OSHA to reconsider removing physicians from a position of responsibility in overseeing all respirator medical qualifications. In any case, for a medical evaluation program to work smoothly, physicians and LHCPs (PLHCPs) must work together as a team. When employers elect to use LHCPs, they should ascertain that the LHCPs have a working relationship with knowledgeable physicians with whom they can work.

Communication

Without clear, timely communication, the medical evaluation will fail to meet participant expectations. At the start of their working relationship, employers and the PLHCP should establish unambiguous guidelines for communications, including their content and timing. The standard requires the employer to provide the PLHCP with a copy of the company's written respiratory protection program, the medical evaluation section of the standard, and specific information regarding the respirator, job demands, and the work environment (Table 2-4). Companies must inform the PLHCP when the working conditions and job demands of respirator users change.

The ultimate responsibility for determining an employee's ability to wear a respirator lies with the employer. However, the employer must consider the PLHCP's medical recommendations. The standard specifies the content of the PLHCP's written recommendation. The recommendation must (1) identify whether or not the worker is medically able to use a respirator, (2) list any limitations on respirator use related to the medical condition of the worker, (3) identify the need, if any, for follow-up medical evaluation, and (4) make a statement acknowledging the worker's receipt of a copy of the recommendation form. If the PLHCP recommends that an employee use a powered air-purifying respirator (PAPR) for medical reasons, the standard requires that the employer provide the PAPR.

On face value, the obligatory communication between the employer and the PLHCP appears simple. In practice, a number of circumstances can delay and impede the transfer of needed information. Frequently, employers do not give providers the essential work-site and respirator information in a timely fashion. The PLHCP may need information from the worker's personal health care provider, may want to conduct further testing, or may refer the worker for follow-up assessment by a specialist.

Table 2-4. Information the Employer Must Provide the Physician or Licensed Health Care Provider

1. Type and weight of the respirator selected for employee
2. Duration and frequency of respirator use (including use for rescue and escape)
3. Expected physical work effort
4. Additional protective clothing and equipment
5. Temperature and humidity extremes that may be encountered
6. Copy of the company's written respiratory program and a copy of CFR1910.134 Sec. (3)

In such cases, medical approval in 24 to 48 hours is unrealistic. Nonetheless, a key indicator of employer satisfaction remains the PLHCP's ability to rapidly complete and communicate medical opinions. Delays affect business operations. Providers must make every effort to expedite the medical evaluation process. They should update the employer with frequent status reports when opinions are delayed because of the need to collect additional information.

Confidentiality

Communication between PLHCPs and employers must not violate professional ethical codes or statutes of confidentiality as defined by OSHA, the Americans with Disabilities Act, and state medical confidentiality laws. OSHA specifically identifies the medical findings and diagnoses as confidential. In the absence of signed medical releases, the sharing of information must be limited to the opinions of the PLHCP regarding a worker's fitness to use a respirator and any necessary restrictions or accommodations.[8] Employers must secure any files containing medical information and keep them separate from all personnel records.[21] PLHCPs and employers should address medical confidentiality at the beginning of their business relationship to clarify expectations of information exchange.

Fit Testing

The standard mandates fit testing for any worker required to wear a tight-fitting facepiece respirator prior to first use and annually thereafter. Fit tests also must occur with changes in the model, size, make, or style of the facepiece or with any change in physical condition of the worker that might affect fit, such as the use of dentures, weight loss, trauma, or other changes in facial structure. OSHA encourages, but does not require, fit testing of workers who wear a respirator voluntarily.

Fit testing aims to ensure that a particular facepiece provides an adequate seal against an individual's face. Several NIOSH studies cited in the standard's preamble document that inattention to the importance of fit testing has been a key factor in overexposing workers to air contaminants.[22] As reviewed in Chapter 1, a respirator's *assigned protection factor* (APF) refers to the maximal level of protectiveness the specific respirator design can achieve under laboratory conditions. Good respirator fit testing helps ensure that an employee achieves protection as close to the APF as possible. *Fit factor* refers to the ratio of contaminants outside versus contaminants inside the facepiece as *actually* measured during an employee's quantitative fit test.

Appendix A of the standard details the mandatory fit-test protocols (Appendix 1A of this book). Based on the type of respirator selected and the exposure conditions for which it will be used, fit testers choose either *qualitative fit testing* (QLFT) or *quantitative fit testing* (QNFT). QLFT may only be used for negative-pressure air-purifying respirators, which must achieve a fit factor of 100 or less, and for all positive-pressure respirators. In contrast, QNFT is an appropriate approach for all classes of respirators (Table 2-5).

Regardless of the fit-testing protocol used, OSHA requires the employer to incorporate certain procedures. To begin, the worker chooses the best-fitting and most comfortable respirator from a selection of models and sizes of the device required for the job. The tester must

Table 2-5. Fit-Testing Guidelines

Respirator type[a]	**QLFT**	**QNFT**
Half-face, negative-pressure APR (<100 fit factor)	Yes	Yes
Full-face, negative-pressure APR (<100 fit factor) used in atmospheres up to 10 times the PEL	Yes	Yes
Full-face, negative-pressure APR (>100 fit factor)	No	Yes
Powered air-purifying (PAPR)	Yes	Yes
Supplied-air respirator (SAR) or self-contained breathing apparatus (SCBA) used in negative-pressure (demand) mode (>100 fit factor)	No	Yes
Supplied-air respirator (SAR) or SCBA used in positive-pressure (pressure demand) mode	Yes	Yes
SCBA, structural firefighting, positive pressure	Yes	Yes
SCBA/SAR, IDLH, positive pressure	Yes	Yes
Mouthpiece and loose-fitting (hood, helmet) respirators	Fit testing not required	

[a]APR = air-purifying; PEL = permissible exposure level; IDLH = immediately dangerous to life and health.

Source: Adapted from *OSHA Respiratory Protection Draft Small Entity Compliance Guide.* Occupational Safety and Health Administration Publication 3079, 1998. Available at *http://www.osha-slc,gov/Publications/secgrev-current.pdf*

instruct the subject in the purpose of the fit test and how to don and adjust the device. While wearing any additional personal protective equipment required by the job, the subject is also asked to perform a user seal check and a series of maneuvers and exercises to ascertain the comfort and acceptability of the facepiece. All respirators are fit tested in the negative-pressure mode regardless of respirator type or fit test used.

The QLFT is a pass/fail test based on the respirator user's ability to detect the test agent. Wearing the chosen respirator, the worker stands in an enclosure while exposed to one of the following OSHA-approved nontoxic test substances. The test begins by ascertaining that the subject has the sensory capability to detect the agent before donning the respirator. Approved test agents are

- Isoamyl acetate (banana oil)—olfactory-dependent
- Saccharin solution or Bitrex (denatonium benzoate) solution—taste-dependent
- Irritant smoke (stannic chloride)—involuntary cough or complaint of irritation

The isoamyl acetate protocol is not appropriate for the fit testing of a particulate respirator unless it is equipped with an organic vapor filter. Because of the potential adverse effects of an irritant on a susceptible worker, NIOSH does not recommend use of the irritant smoke approach.[20] During the test, the subject performs a variety of exercises that challenge the face seal of the respirator, such as head movements, deep breathing, talking, and jogging in place. A failed test occurs anytime the respirator user identifies the test agent.[9]

Quantitative testing (i.e., QNFT) differs from the qualitative testing (i.e., QLFT) in that it relies on measurement of leakage into the respirator of the test agent to ascertain the acceptability of the facepiece. The QNFT method entails placement of a sampling probe through the facepiece. The respirator user enters the test chamber into which a nontoxic agent is aerosolized. The tester samples the concentration of the aerosol inside the enclosure and inside the respirator. The QNFT calculates the ratio of the aerosol concentration inside the test chamber to the aerosol concentration inside the respirator. The QNFT protocol considers the fit acceptable if

- For a half- or quarter-facepiece device, the ratio (fit factor) achieved is greater than or equal to 100.

- For a full-facepiece device, the ratio is greater than or equal to 500.[9]

Respirator Use

OSHA requires that employers establish and implement the following procedures to ensure that workers use respirators appropriately:

1. *Protection of facepiece seal* (required only for tight-fitting respirators). Both the employer and the worker have the responsibility to identify and mitigate any factors that interfere with facepiece seal. Relevant factors that may change after the initial fit testing include facial hair between the face and the sealing surface, facial scars, dentition, jewelry, corrective lens frames, or other PPE. Prior to use, workers should verify seal and proper operation of the respirator with a "user seal check" using the procedures outlined in Appendix B1 of the standard (Appendix 1B1 of this book) or those recommended by the respirator manufacturer.[9]

2. *Assurance of continuing respirator effectiveness throughout work shifts.* Companies must establish procedures to ensure respirator effectiveness throughout a work shift. Ongoing surveillance of working conditions and employee exposures detects changes that may necessitate a reevaluation of the effectiveness of selected respirators. Employers must replace respirators of their air-purifying elements according to established change-out schedules, warning ESLIs, or whenever the employee detects breakthrough or malfunctioning of the device. Employers also must take steps to prevent removal of respirators in hazardous environments. They should require workers to leave the respirator use area when

- Facepiece adjustment or washing is necessary to maintain comfort.
- Breakthrough or other problems with respirator function are noted.
- The respirator or its air-purifying element must be replaced.

3. *Establishment of procedures for use of respirators in an IDLH environment.* This section of the standard requires standby personnel to be ready to assist respirator users working in an IDLH environment. The responsibilities of standby personnel include maintaining communication (voice, visual, or hand signals) with the respirator user

and rescue if the worker's respirator fails. Employers must equip standby personnel with appropriate OSHA-designated respirators and rescue equipment.[9]

4. *Establishment of procedures for interior structural firefighting.* For interior structural firefighting, OSHA adds further requirements to the provisions mandated in an IDLH atmosphere. The standard requires that two employees stand ready outside the IDLH environment to rescue the worker. All employees must wear SCBAs for this work.

Maintenance and Care of Respirators

Without care and maintenance, personal respiratory protective devices will fail. The OSHA standard requires that companies identify procedures that address each of the following:

- Cleaning and disinfecting of respirators
- Appropriate storage to prevent respirator degradation
- Frequency, method, and documentation of respirator inspections
- Removal and repair of defective respirators

Atmosphere-Supplying Respirators

Atmosphere-supplying respirators provide a lifeline for employees working in hazardous atmospheric conditions. OSHA requires that employers meet multiple specifications to ensure the purity and safety of the respirator air supply. Compressed breathing air, which is used most commonly, must meet the grade D requirements defined in the ANSI *Compressed Gas Association Commodity Specification for Air* (G-7.1-1989). When air is supplied by a vendor, the employer can rely on the vendor's certificate of meeting the required specifications.[9] Additional employer responsibilities regarding atmosphere-supplying respirators include ensuring that cylinders meet standard requirements, monitoring of carbon monoxide, and prevention of contamination of supplied air. Couplings used on the respirator air line must not be compatible with other nonbreathable work-site air and gas system outlets.

Employee Training

All employees required to wear a respirator must take part in the company's respiratory protection training program prior to respirator use, annually, and at any time there are changes in the workplace or respirator type. Employees who use a respirator voluntarily need not attend a training program but must receive a copy of the standard's Appendix D (Appendix 1D of this book), entitled "Information for Employees Using Respirators when Not Required under the Standard." At the conclusion of the training program, employees must demonstrate knowledge in the following areas:

- Why the use of a respirator is necessary in their work environment
- How improper fit, use, and maintenance can make the respirator ineffective
- Limitations and capabilities of the selected respirator
- Effective use of a respirator in emergency situations, including respirator malfunction
- Inspection, donning, and removal of the respirator
- User seal check of the respirator
- Maintenance and storage procedures
- Medical signs and symptoms that may limit or contraindicate use of a respirator
- Employer requirements under the standard

Employees who demonstrate a lack of knowledge, inappropriate use, or improper care of a respirator require retraining.[9]

Program Evaluation

OSHA requires that employers evaluate the effectiveness of the respiratory protection program by consultation with employees and workplace audits to ensure correct implementation of all provisions of the written respiratory protection program. Companies may determine the frequency of program evaluation with reference to the complexity of the program and the number of respirator workers. OSHA requires that areas for program review include at least appropriate respirator selection, fit, use, and maintenance. Program reviewers should document problems identified and corrective actions.[9]

Recordkeeping

OSHA mandates that employers maintain certain records pertaining to their respiratory protection program. Recordkeeping helps employers audit their program, it facilitates employee involvement, and it provides a compliance record for OSHA. Records should include the written respiratory protection program, fit-test records, and medical evaluation records. Fit-test records must include employee name, type of test performed and results, the make, model, and size of the respirator, and the date of the test. Records are maintained until the individual's next fit test. Medical records must be maintained in accordance with OSHA's access to employee exposure and medical records standard (29 CFR 1910.1020).[9]

Conclusion

The comprehensive systems approach of the standard aims to ensure the health and safety of respirator workers. The complexity of an employer's program depends on the number and type of occupational exposures for which respirators must provide protection. To perform their role optimally, health care providers must have a general knowledge of an employer's respiratory protection program.

Further Reading

OSHA's Web site provides a rich lode of easily accessible additional information about the revised respiratory protection standard. These are the most useful sections for occupational health care professionals interested in assisting their clients with OSHA compliance.

Respiratory Protection Advisor. Available at http://www.osha-slc.gov/SLTC/respiratory_advisor/mainpage.html. This advisor provides detailed information about the proper selection of respiratory protection and the development of change schedules for gas/vapor cartridges. Links from this page take the viewer to all of OSHA's technical guidance related to the revised respiratory protection standard.

OSHA Preambles: Respiratory Protection. Table of Contents by Section. Available at *http://www.osha-slc.gov/Preamble/RP_toc/RespiratoryProtection_toc_by_sect.html. This page provides links to all of the OSHA preambles. The preambles offer fascinating reading about how the standard reached its current form. In particular, the rationale for each section of the standard is detailed.*

Draft Small Entity Compliance Guide for Respiratory Protection Standard. Available at *http://www.osha-slc.gov/Publications/secgrev-current.pdf. This site lists questions and answers, provides checklists for compliance, and gives an example of a company' respiratory protection written program.*

OSHA Questions and Answers on the Respiratory Protection Program. Available at *http://www.osha-slc.gov/Publications/secgrev-current.pdf. This site lists and answers many of the most common questions that employers or health care providers may have about administration of the revised respiratory protection standard.*

OSHA Standards Interpretation and Compliance Letters. Table of Contents for 29 CFR Standard 1910.134. Available at *http://www.osha-slc.gov/OshDoc/Interp_Std_toc/Interp_TOC_for_1910_0134.html?Interpretation&28s &29=Interpretation%28s%29. This site provides links to OSHA letters of interpretation about specific details of the respiratory protection standard.*

References

1. Harber P, Merz B, Chi K. Decision model for optimizing respirator protection. *J Occup Environ Med* 1999; 41:356–365.
2. Que Hee SS, Lawrence P. Inhalation exposure of lead in brass foundry workers: The evaluation of the effectiveness of the powered air-purifying respirator and engineering controls. *Am Ind Hyg Assoc J* 1983; 44:746–751.
3. Brousseau LM, Traubel K. An evaluation of respirator maintenance requirements. *Am Ind Hyg Assoc J* 1997; 58:242–246.
4. OSHA Preambles: Respiratory Protection Standard. III. Events Leading to the Final Standard. Available at *http://www.osha-slc.gov/Preamble/RP_html/RESPIRATORY3.html.*
5. OSHA Standards cited for all SICs; all sizes. Available at *http://www.osha.gov/cgi-bin/std/stdser1?esize=0&state=FEFederal&sic=All.*
6. OSHA Improves Respirator Protection for 5 Million Workers in 1.3 Million Worksites. U.S. Department of Labor, Office of Public Affairs, OSHA National News Release USDL 98-04, January 8, 1998. Available at *http://www.osha.gov/media/oshnews/jan98/osha9804.html.*
7. Fairfax RE. Preemption of OSHA's respiratory protection standard by DOT's Office of Pipeline Safety regulations. OSHA Standards Interpretation and Compliance Letters, April 8, 1999. Available at *http://www.osha-slc.gov/OshDoc/Interp_data/I19990408.html.*
8. Questions and Answers on the Respiratory Protection Standard. Available at *http://www.osha.gov/qna.pdf.*
9. Small Entity Compliance Guide for the Revised Respiratory Protection Standard. Available at *http://www.osha-slc.gov/Publications/secgrev-current.pdf.*

10. OSHA Directives. CPL2-0.120. Inspection Procedures for the Respiratory Protection Standard. Available at *http://www.osha-slc.gov/OshDoc/Directive_data/CPL_2-0_120.html.*
11. Harber P. Respirators. In Rom W (ed.), *Environmental and Occupational Medicine,* 3d ed. Philadelphia: Lippincott-Raven, 1998:1757–1765.
12. *NIOSH Respirator Decision Logic.* Publication 87-108. Washington: National Institute of Occupational Safety and Health, 1987.
13. Industrial Safety Equipment Association. Use *and Selection Guide for Non-Powered Air Purifying Particulate Respirators.* Arlington, VA: ISEA, 1996.
14. Occupational Safety and Health Administration. Assigned Protection Factors. Reprinted from a table available at *http://www.osha-slc.gov/SLTC/respiratory_advisor/haz_expose/apf/apf.html.*
15. Fairfax RE. Medical evaluation not required for use of escape only respirators. OSHA Standards Interpretation and Compliance Letters, March 8, 1999. Available at *http://www.osha-slc.gov/OshDoc/Interp_data/I19990308.html.*
16. Fairfax RE. Employee declination of medical evaluation; religious exemptions for respirator use. OSHA Standards Interpretation and Compliance Letters, December 2, 1998. Available at *http://www.osha-slc.gov/OshDoc/Interp_data/I19981202.html.*
17. Fairfax RE. Clarification of the medical evaluation provisions of the revised respiratory protection standard. OSHA Standards Interpretation and Compliance Letters, November 16, 1998. Available at *http://www.osha-slc.gov/OshDoc/Interp_data/I19981116.html.*
18. Fairfax RE. Medical evaluation requirements under the respiratory protection standard. OSHA Standards Interpretation and Compliance Letters, October 5, 1998. Available at *http://www.osha-slc.gov/OshDoc/Interp_data/I19981005.html.*
19. Fairfax RE. Respiratory protection compliance responsibility for contract workers and visitors. OSHA Standards Interpretation and Compliance Letters, September 15, 1987. Available at *http://www.osha-slc.gov/OshDoc/Interp_data/I19870915.html.*
20. Rosenstock L. *NIOSH Respirator Use Policy.* Washington: NIOSH, August 4, 1999.
21. Langer C. Title I of the Americans with Disabilities Act. *Occup Med* 1996; 11:5–16.
22. OSHA Preambles: Respiratory Protection Standard. Section VII. Summary and Explanation Available at: *http://www.osha-slc.gov/Preamble/RP_html/RESPIRATORY7.html.*

3

Medical Evaluation Procedures

The medical evaluation for respirator use takes place in a continuum of events that culminate in a medical opinion about an individual's capacity to safely work while wearing a respirator (Figure 3-1). This chapter describes both the content and administrative mechanisms appropriate for collecting clinical data. The original respiratory protection standard mandated that a physician medically qualify workers to wear a respirator prior to first use and periodically thereafter. The standard did not set out the features of the evaluation. In contrast, the revised standard provides specific guidelines for the administration, minimum content, and communication of the medical evaluation. These changes also address business concerns such as the cost of the evaluation.

Occupational Safety and Health Administration (OSHA) standards undergo analysis by the Department of Labor, Office of Management and Budget, and the Small Business Administration to assess their financial impact on business and industry. OSHA predicts that the revised respiratory protection standard will cost no more than 0.02 percent of small business revenues to implement.[1] To minimize the financial burden to employers, OSHA enacted a number of cost-saving measures in the revised standard. Several of these measures involve the medical evaluation: the option for a medical questionnaire in lieu of a traditional clinical examination, the choice of using a nonphysician licensed health care provider (LHCP) to conduct the evaluation, and elimination of routine periodic reevaluation. Before performing a medical evaluation for respirator use, the physician or LHCP should ascertain that the employer has conducted a hazard assessment and selected respirators for use (see Table 2-2). The employer may concisely convey workplace and respirator information to the PLHCP using a form such as the Environmental Conditions and Respirator Use Form (see Appendix 2). Employers with multiple respirator use scenarios must supply information appropriate to the specific worker undergoing medical evaluation. The employer and PLHCP should agree on the evaluation process, communication expectations, and costs. If the

Figure 3-1. The Medical Evaluation for Respirator Use*

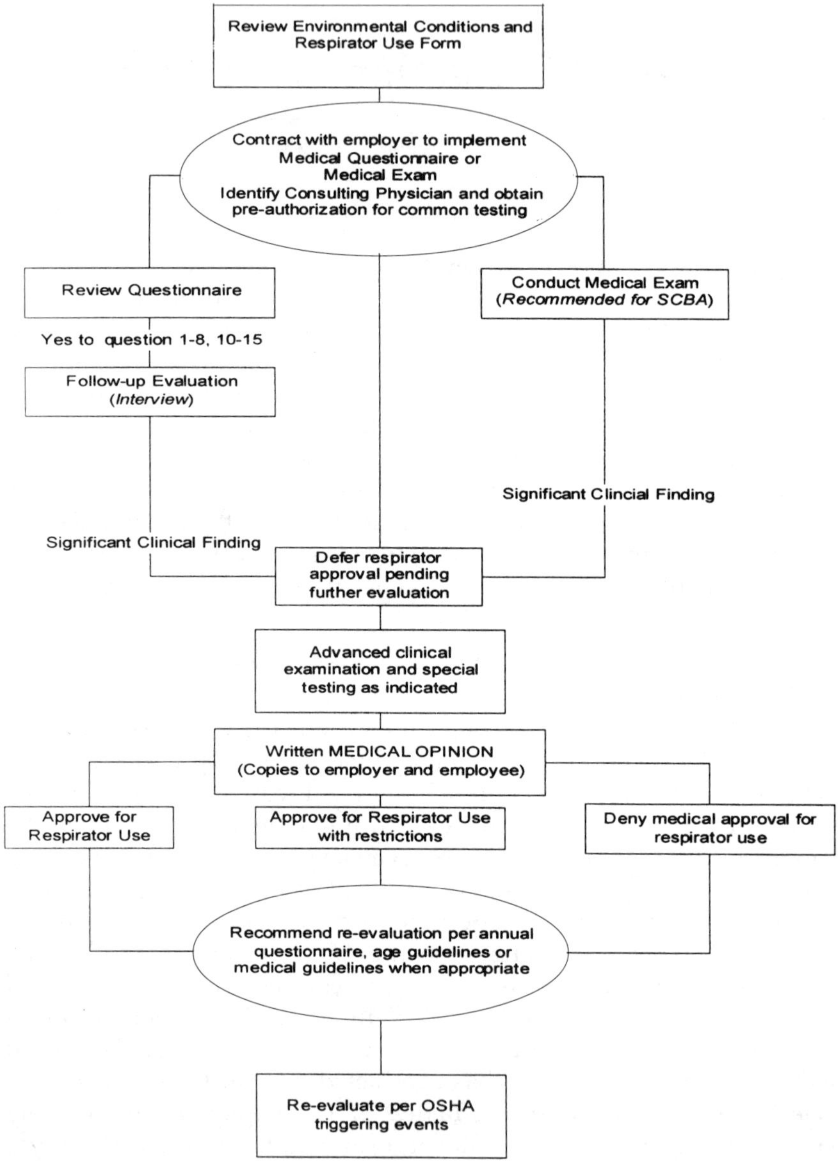

*Ellipses and italics designate activities recommended by authors. All other text and rectangles are OSHA mandated.

employer has chosen a LHCP to conduct the evaluation, both parties also should agree on a specific physician to clinically support the LHCP as needed. This physician should have expertise in occupational medicine and skills and experience in the evaluation of workers using respirators. Although primary care and specialist physicians can assist in defining the nature of a worker's health problems and in optimizing their treatment, they rarely have the requisite understanding of job demands or the psychophysiologic burden of respirator use. Since at least some follow-up testing usually is required, contracts should include authorization for the PLHCP to proceed with commonly required testing without consulting the employer.

29 CFR 1910.134 requires each company to establish procedures for confidential administration of the medical evaluation during an employee's normal working hours or at a time and place convenient to the employee. In addition, the employee must be provided with an opportunity to discuss the questionnaire and the results of any examination with the PLHCP. Neither completed questionnaires nor medical examination records can be maintained by employers, except when an employee health office exists that is administratively separate from the company's central administration.[2] The confidentiality of these records must be ensured by establishing medical access protocols as mandated by 29 CFR 1910.1020.

OSHA's Respirator Medical Evaluation Questionnaire

The medical questionnaire serves as a screening tool to identify workers requiring further medical evaluation prior to respirator use. The respirator medical evaluation questionnaire, found in Appendix C of the standard (Appendix 1C of this book) consists of two parts. Part A contains mandatory questions that cannot be altered. Part B is voluntary; its use is at the discretion of the PLHCP. Section 1 of Part A of the questionnaire identifies worker demographics, respirator use data, and PLHCP contact information. Section 2 of Part A has two distinct portions. All prospective respirator workers must answer questions 1 through 9. These questions screen the worker for evidence of a variety of conditions relevant to respirator use, including diabetes; allergies; trouble smelling odors; cardiovascular, respiratory, seizure, or anxiety disorders; and problems with past respirator use. Employees whose jobs require them to wear a full-facepiece respirator or a self-contained breathing apparatus (SCBA) also must answer questions 10 to 15. These questions explore problems that may be exacerbated by heavy and bulky respirators, including sensory impairment and musculoskeletal difficulties.

Questionnaire Administration

The task of administering the questionnaire rests with the employer's program administrator or representative. Employers must ensure that each employee understands the content of the questionnaire. Illiterate employees and those who cannot read English must be given assistance. In such cases, the employer may ask the employee if a friend or relative can provide the needed assistance. OSHA does not require that the employer hire an interpreter.[2] However, Spanish translations of the form are available (see Appendix 1C). Alternatively, the employer may elect to bypass the questionnaire and refer the individual directly to the PLCHP, who can assist the employee in filling out the questionnaire (with an interpreter, if necessary) or simply conduct the medical evaluation.

The employer's program administrator also must develop procedures that allow confidential completion of the questionnaire and its transfer to the PLHCP. Confidentiality also extends to any PLHCP telephone or face-to-face contact with the worker. Either the employer's program administrator or PLHCP may distribute the questionnaire to workers, but the employer may not have access to an individual's answers to the questions. If the questionnaire is provided at the work site, employers must afford individuals privacy to answer the questionnaire, such as in a medical treatment area, a conference room, or another area separate from the work area. Companies with an onsite occupational health clinic can ask employees to deliver their questionnaires directly to the clinic. If there is no onsite clinic, employees can place the questionnaires in a locked strongbox with a mail slot. Off site PLHCPs then can travel to the workplace to collect and review questionnaires placed in a strongbox by the employee. Alternatively, workers can use preaddressed stamped envelopes to mail completed questionnaires to offsite health care providers.

Questionnaire Review

We suggest that PLHCPs review the questionnaires with reference to the guidelines in Chapter 4. In the absence of any positive responses to the questions in Part A, Section 2, the PLHCP can approve the examinee for respirator use without further medical investigation. According to the standard, a single affirmative response to any question except question 9 in this section requires further medical evaluation. OSHA gives the PLHCP the responsibility to define the content of the "further medical evaluation." It may include a telephone conversation between the PHLCP and the worker, communication with the worker's treating physician(s), or a follow-up medical examination and laboratory testing.[2,3] Questionnaires with answers that trigger the requirement for follow-up fall into two categories: (1) those for which an interview alone quickly resolves any concerns about respirator use, and (2) those which require communication with attending physician(s) or further, more complex medical evaluation.

The Medical Examination, Administration, and Review

The traditional medical examination approach to qualify a worker for respirator use must at a minimum gather the information requested in the OSHA medical questionnaire. The actual questionnaire does not have to be used. Additional data collection is at the discretion of the examiner. Usually the PLHCP performs an interview, physical examination, and often spirometry.[4] The revised standard provides the rationale to narrow the investigation from a complete history and physical examination to an evaluation that targets relevant body systems: cardiovascular, endocrine, facial configuration, hearing, musculoskeletal, neurologic, olfactory, psychological, respiratory, and vision (see Appendix 3). Most important, the interviewer should inquire about past respirator use and tolerance.[5]

Prior to the examination, we recommend that workers sign a consent for evaluation form (see Appendix 4). LHCPs may conduct all aspects of the examination within their scope of practice. Note, however, that scope of practice varies with each state.

OSHA leaves interpretation of the significance of any findings obtained of a medical examination up to the PLHCP (see Chapter 4). In the absence of significant relevant findings, the PLHCP may medically qualify the examinee for respirator use. Significant findings may allow a quick decision to disqualify the worker or restrict respirator wear or specific tasks. More commonly, significant findings require further evaluation before a final determination can be made. As with the questionnaire, the PLHCP defines the nature of any further evaluation.

Choosing an Approach to the Medical Evaluation

OSHA relied on several studies to bolster its decision to offer the medical questionnaire option. Few workers (between 2 and 2.5 percent) receive medical restrictions or disqualification for respirator use.[6,7] One study, involving over 400 workers, demonstrated that a questionnaire had 100 percent sensitivity in identifying workers requiring work restrictions when compared with the recommendations based on a medical examination. However, the same study, which used a questionnaire similar to OSHA's, found that 81 percent of participants required some further medical evaluation. Of the group that received further evaluation, 97 percent were given medical clearance for respirator use.[8] The American Thoracic Society also considers a screening questionnaire an appropriate method for the medical evaluation of workers required to wear respirators.[9]

Use of a questionnaire holds the promise of reduced examination costs, decreased losses in productivity, and quicker completion.[7] The standardized questionnaire also ensures that all worker evaluations are consistent, regardless of changes in health care personnel. Workers benefit from a less intrusive, more focused evaluation.

Employers should recognize the limitations inherent in a questionnaire as well as certain hidden costs. The questionnaire approach requires overcoming language and literacy barriers. A worker embarrassed by illiteracy or a language barrier may answer questions without understanding their meaning. Questionnaire administration at a workplace without an onsite health service raises confidentiality issues. Moreover, a questionnaire cannot detect asymptomatic health problems such as hypertension. It also makes it easy for a worker to conceal or inadvertently not reveal medical conditions. When the offsite PLHCP reviews a questionnaire, the practical issues involved in setting up follow-up interviews may add considerable time and frustration to the approval process. The high rate of follow-up testing prompted by the questionnaire used in the previously mentioned study also suggests that the hope of reduced costs and quicker completion may be false.[8]

The medical examination has its own strengths and weaknesses. It allows for a more comprehensive health assessment. For instance, the only neurologic condition the questionnaire addresses is seizure disorder. There are no questions that address loss of consciousness, dizziness, or syncope, yet any of these problems and others may interfere with the safe use of a respirator. Conducting a medical examination also may uncover underlying disease unknown to the worker, such as hypertension, which may benefit from treatment. The examination approach also consolidates the interview and physical examination into a one step, potentially decreasing the time needed to complete the evaluation. On the other hand, a medical examination may be unnecessary in the majority of workers.[6]

Besides subjecting workers to unnecessary invasion of their privacy, the cost of a medical examination, combined with time away from work, exceeds that of a questionnaire. Companies with an onsite health service or that contract with PLHCPs to conduct examinations onsite may minimize the expense of lost labor time. The expense of sponsoring onsite health services may counterbalance any savings accrued by reducing hours of lost labor. The most cost-effective approach to conducting respirator qualification examinations will depend on such factors as the number of examinations, the distance of occupational medical facilities from the company, and the range of occupational health services that an onsite program may offer.

No specific formula exists for determining which medical evaluation approach is most appropriate for an employer. Information about the demographics of a company's workers, the intensity of job demands, additional necessary personal protective equipment (PPE), the work environment, and the types of respirators should help the employer choose the most cost-efficient and medically appropriate approach to evaluation. The older the working population and the higher the percentage of smokers, the greater is the number of follow-up evaluations the OSHA questionnaire triggers. The time and professional cost of these additional evaluations may overwhelm any initial cost savings in choosing the questionnaire approach. An employer may lean toward a hands-on clinical examination by skilled practitioners as the risk of an adverse health or safety event increases. We recommend that all workers wearing SCBAs undergo a medical evaluation. Conversely, situations involving sedentary work with minimal environmental risk and a generally young work force may best be served by the medical questionnaire option. Table 3-1 summarizes the pros and cons of the different approaches to medical evaluation.

Follow-up Medical Evaluation

Concerns identified by either the medical examination or the questionnaire trigger the follow-up medical evaluation. The follow-up medical evaluation may or may not result in a new diagnosis but always should gather more objective information about the capability of the worker to wear the designated respirator. The process may include further history and physical examination, medical testing, work trials, and consultation with the worker's health care provider(s). The OSHA respiratory protection standard requires the employer to "pay for the medical evaluation and any related expense incurred by the employee during the evaluation, including travel costs."[2] Prior to beginning their working relationship, the PLHCP should establish likely additional evaluation costs with the employer. Before incurring unanticipated costs, the PLHCP should consult with the employer.

Table 3-1. Strengths and Weaknesses of Medical Questionnaire versus Traditional Medical Examination

Strengths	Weaknesses
Medical Questionnaire	
• Less intrusive • Low cost • Self-administered • Focused and standardized • Minimizes unnecessary medical testing • Little time required to complete • Initial review simple	• Literacy and language barriers • In absence of onsite provider, additional confidentiality concerns • Relies on subjective responses • Cannot detect undiagnosed, asymptomatic conditions • High rate of follow-up delays approval process
Traditional Medical Examination	
• More comprehensive evaluation • May identify new, treatable medical problems • History, physical, and recommendation as one-step process • Objective information obtained • Not as language-dependent	• More intrusive • Often adds little useful information • Higher cost

Medical Reevaluation

OSHA no longer requires routine periodic, time-based evaluations. Instead, reevaluations are event-based. Paragraph (e)(7) of the standard specifies these events as follows: when the employee reports signs or symptoms that are relevant to the employee's ability to use a respirator; when a PLHCP, supervisor, or respirator program administrator informs the employer that an employee needs to be reevaluated; when information from the respirator program, including observations made during fit testing or program evaluation, indicates a need for employee reevaluation; or when a change in workplace conditions occurs that may result in a substantial increase in the physiologic burden that respirator use places on the employee (Table 3-2). Paragraph (e)(7) thus provides the PLHCP with authority to recommend additional, periodic evaluations based on the worker's medical condition or other considerations. The Nuclear Regulatory Commission places the burden on the physician to establish a schedule of additional examinations for radiation workers.[10] Many employees who use a respirator are also in respiratory surveillance programs that involve annual reevaluation to monitor the adverse effects of specific occupational exposures

OSHA chose to forgo time-based intervals for respiratory evaluations to minimize the regulatory burden. The agency concluded that event-based reevaluations would identify most individuals whose respirator use might benefit from modification. Evidence indeed suggests that routine, age-based guidelines for medical reevaluation probably do not meet reasonable criteria of cost-effectiveness.[11] On the other hand, given that violations of the respiratory protection standard remain among the top 10 most frequent OSHA findings, employers cannot be assumed to reliably trigger timely medical reassessments. As with all OSHA standards, the mandates dictate the minimum acceptable practice. We suggest two ways for PLHCPs to improve the likelihood of appropriate reevaluation in keeping with OSHA's mandated triggers. First, health care providers should take advantage of the standard's provision that allows them to recommend reevaluation based on the medical condition of the employee. Chapter 4 provides condition-specific guidelines for the frequency of this type of reevaluation. Second, we recommend that prior to the mandated annual fit test, employees complete a brief questionnaire. This questionnaire identifies any relevant changes in a worker's health or respirator tolerance that warrant PLHCP evaluation. As illustrated in Table 3-3, this questionnaire does not ask for any medical information requiring confidentiality. More simply, it asks employees if, from their perspective, any of the potentially

Table 3-2. OSHA Triggers for Medical Reevaluation

1. Employee reports medical signs or symptoms related to respirator use.
2. Physician or licensed health care professional, supervisor, or program administrator informs employer that employee needs to be reevaluated.
3. Information from the respiratory protection program, including observations made during fit testing and program evaluation, indicates a need for employee reevaluation.
4. Change occurs in working conditions that may result in a substantial increase in the physiologic burden placed on an employee.

triggering events described by OSHA have occurred. If an employee answers any of the questions affirmatively, the fit tester should refer the individual to a PLHCP for reevaluation.

The OSHA standard does not prohibit employers from instituting a time-based reevaluation in addition to the required event-based triggers. The American National Standards Institute's ANSI Z88.6-1984,[12] The National Institute for Occupational Safety and Health's *NIOSH Respirator Decision Logic,*[13] and the American College of Occupational and Environmental Medicine's (ACOEM's) Occupational and Environmental Medical Practice Committee[14] provide PLHCPs with a rationale to suggest additional medical evaluations at time-based intervals based on worker age and the type of respirator used. Each of these organizations advocate that employers screen older employees more frequently than younger employees because of the heightened

Table 3-3. Questionnaire Prior to Annual Fit Testing

Name: ________________________ DOB: ____________

Department: ____________________ Job: ________________

1. Have you developed any medical problems or symptoms that may limit your ability to use a respirator?

   [ ] No [ ] Yes

2. Have you been told by a health care professional, your supervisor, or the respirator program administrator that you should be medically reevaluated?

   [ ] No [ ] Yes

3. Has there been a change in workplace conditions, e.g., physical work effort, protective clothing, temperature, that has resulted in a substantial increase in the physical burden on you?

   [ ] No [ ] Yes

It is your responsibility to report to your supervisor any change in health status that may affect your ability to use a respirator.

Employee signature: ____________________ Date: ____________

[ ] No medical follow-up is necessary; proceed with fit test.

[ ] Medical evaluation is indicated; do not proceed with fit test

Reviewer signature: ____________________ Date: ____________

risk of cardiovascular and respiratory disease associated with age (see Table 3-4).

Commonly Performed Tests as Part of the Evaluation of Respirator Users

Spirometry

Since respirators place an additional burden on ventilation, examiners frequently perform spirometry to assess the lung function of workers using respirators. The PLHCP should not rely on test results as the sole determinant of respirator tolerance by a worker, however. The American Thoracic Society (ATS) notes that "pulmonary function tests have not been demonstrated to provide sensitive or specific indicators for respirator intolerance or safety."[15] On the other hand, the ATS considers spirometry useful in assessing a worker's fitness to cope with specific job demands. Furthermore, the ATS identifies certain respirator worker populations as particularly appropriate for spirometry testing. These include workers who report respiratory symptoms, workers older than age 45 wearing SCBAs under strenuous conditions, and workers 55 years of age or older.[9]

Spirometric indices considered appropriate for the evaluation of the respirator worker include the forced vital capacity (FVC), the forced expiratory volume in 1 second (FEV_1), and the FEV_1/FVC ratio. The ATS discourages use of the forced expiratory flow from 25% to 75% of vital capacity (FEF_{25-75}) or other measures of instantaneous flow to diagnose obstruction. Keeping the number of tests to a minimum, as recommended by the ATS, limits the number of false-positive tests.[15] Once touted as a good predictor of respirator tolerance, the maximum

Table 3-4. Age-Based Recommended Medical Evaluation Frequency

	Frequency	
Age	**Light to moderate work**	**Strenuous work with SCBA**
<35	Every 5 years	Every 3 years
35–45	Every 2 years	Every 18 months
>45	Every year	Every year

voluntary ventilation (MVV) is no longer recommended for this purpose.[16] Depending on their state-specific scope of practice and following appropriate clinical guidelines, LHCPs may initiate spirometry testing. We recommend that interpretation of test results and their impact on the recommendation for respirator use be left to the physician or midlevel clinician (nurse-practitioner or physician's assistant).

Selecting Reference Ranges

The interpreter of a spirogram compares FVC, FEV_1, and FEV_1/FVC measurements with average values "predicted" for the subject based on a reference population. Regression equations used in the automated calculations of most modern spirometers summarize relationships between sex, age, height, race, and pulmonary function to determine the predicted value for a particular subject. Choosing the reference population that best matches the employee will yield the most accurate results. In clinical occupational settings, the Knudson data set and equations are widely used because they were incorporated into the OSHA cotton dust standard when it was first promulgated. Different prediction equations lead to varying "normal ranges." The ATS suggests empirically checking the fit of a set of reference values to a particular occupational setting by testing 20 to 40 local nonsmoking healthy subjects and determining their percentages of predicted values using the intended reference equations.[16]

Knudson and other Caucasian data sets in current clinical use in North America do not adequately represent the racial diversity of the U.S. population, let alone the world. To account for observed racial differences in pulmonary function, OSHA's cotton dust standard mandated that a factor of 0.85 be used to adjust Knudson predictive values for African-Americans.[18] An important new set of population spirometric data from the third National Health and Nutrition Examination Survey (NHANES III) became available in January of 1999.[19] The NHANES III data calculate reference equations separately for Caucasians, African-Americans, and Hispanics. ACOEM recommends that these data and regression equations be used in the future as they become available in automated spirometers. Until these equations are available, ACOEM recommends that predicted values for Caucasians be race-adjusted for African-American, Chinese, and Japanese subjects, applying the ATS recommended adjustment of 0.88, if possible, or else 0.85, to the Caucasian predicted FEV_1 and FVC.[20] Technicians should rely on test subjects' racial self-identification. When practicing in Europe, clinicians should employ European reference values.[15]

Interpretation

Interpretation of spirometry begins with an assessment of its quality (see Appendix 5). Once confident with the quality of the spirogram, the clinician should use the interpretive algorithm provided by ACOEM[21] (Table 3-5). In keeping with the standard clinical practice of interpreting clinical laboratory information, the ACOEM algorithm guides the interpreter to use lower limits of normal (LLN) values that represent the 5th percentile of the reference population rather than the popular (and easy to remember) 80 percent of predicted value. The LLN value must be obtained from tables or equations published with the reference population data used or calculated as LLN = 1.645 SEE (standard error the estimate).[19,22] (See Table 3-6 for sample LLN values for a working population.) Use of LLN values helps to avoid false-positive interpretations.

The ACOEM algorithm directs the interpreter to begin by assessing the FEV_1/FVC ratio. When this is below the LLN value, the clinician then looks at the FEV_1. If the FEV_1 is also below the LLN value, airway obstruction is diagnosed. If the FEV_1 is normal in the presence of a low FEV_1/FVC ratio, borderline obstruction is diagnosed. If the FEV_1/FVC ratio is normal or high, the interpreter reviews the FVC. If the FVC is below the lower limit of a subject's normal range, pulmonary restriction is suggested. The amount of reduction in flow rates determines the severity of the obstructive or restrictive impairments (see Table 3-5).

The technology of computerized spirometers has reduced the variability of specific individual spirometric values resulting from instrumentation.[23] Nonetheless, spirometric results remain subject to a number of variables that influence the validity and reliability of the test, including coaching technique, subject effort, body posture (sitting or standing), time of day, and environmental conditions.

Conclusion

Spirometry can provide data that are useful in assessing a worker's ability to work safely while wearing a respirator. However, this information will be useless, misleading, and perhaps harmful unless it is collected and interpreted with careful attention to ATS and ACOEM guidelines.[15,20]

Table 3-5. Spirometry Interpretation Flowchart

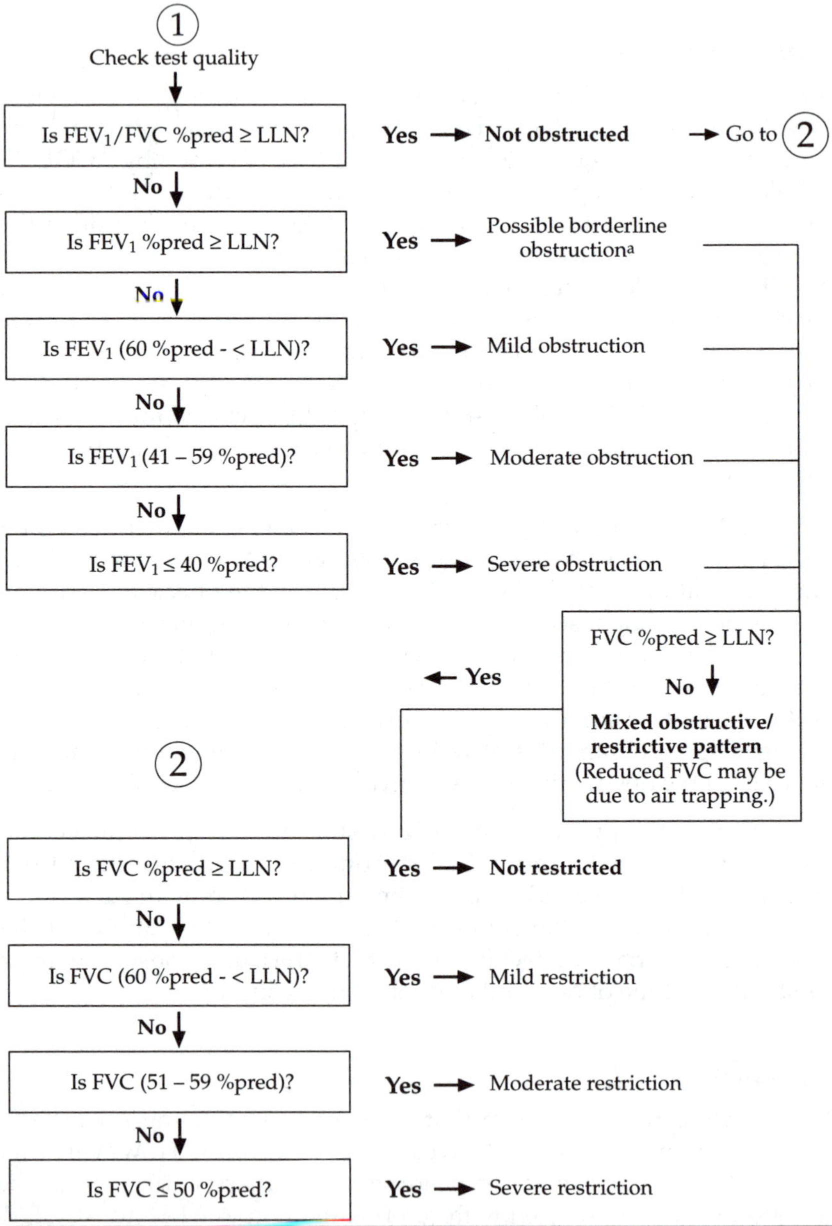

[a]See text.

Source: ACOEM. Spirometry in the Occupational Setting. Available at http://www.acoem.org/paprguid/guides/spaper/spirom.htm

Table 3-6. Lower Limit of Normal for Spirometric Percent Predicted Values from Knudson Prediction Equations

Age	FVC	FEV_1	$FEV_1/FVC\%$	FEF_{25-75}
		Females		
25–39	76.9%	70.3%	85.9%	55.3%
40–69	75.2%	77.9%	85.9%	59.2%
		Males		
25–39	81.1%	79.1%	86.9%	55.3%
40–69	73.4%	77.2%	86.9%	40.3%

Source: Adapted from Knudson RJ, Lebowitz MD, Holberg CJ, et al. Changes in the normal maximal expiratory flow-volume curve with growth and aging. *Am Rev Respir* Dis 1983; 127:725–734.

Chest X-rays

Routine, screening chest x-rays have no place in the respirator approval process unless they are obtained in conjunction with a medical surveillance evaluation such as mandated by OSHA's asbestos standard. On the other hand, a physician or midlevel clinician may appropriately order a chest x-ray for diagnostic evaluation when it is indicated by clinical findings. The significance of an individual's chest x-ray results for respirator use must be determined by a physician or midlevel clinician, with consultation as needed.

Electrocardiography

Resting electrocardiography (ECG) and exercise ECG are among the most common laboratory tests used for the evaluation of cardiac disease. Although they play a clear role in the diagnosis of cardiac problems, their utility in the evaluation of asymptomatic coronary artery disease (CAD) has been questioned. In fact, the U.S. Preventive Services Task Force concludes that there is insufficient evidence to recommend for or against screening middle-aged and older men and women for asymptomatic CAD using resting ECG, ambulatory ECG, or exercise ECG. The task force recommends against routine screening as part of periodic health visits or preparticipation sports examinations for young adults.[24] On the other hand, the task force does recommend screening individuals in certain occupations where the risk of sudden death or incapacitation would endanger the safety of others. The task force notes that although screening cannot reliably identify all persons at risk of an acute event, it may increase the margin of safety in the workplace.

The American Academy of Family Practice recommends exercise ECG for those whose jobs are linked to public safety and require high cardiovascular performance.[25] In addition, certain federal agencies such as the Department of Transportation, the Nuclear Regulatory Commission, and the Federal Aviation Authority recommend screening certain classes of workers for asymptomatic heart disease. ANSI considers a cardiac exercise stress test as a reasonable recommendation for personnel who use an SCBA or a rebreather respirator in strenuous work conditions.

We concur with the American College of Sports Medicine (ACSM) recommendations for risk stratification as an approach to designing the appropriate level of health screening prior to initiating an exercise regime. The ACSM suggests categorizing individuals as "apparently healthy," "increased risk" (one or more cardiopulmonary signs or symptoms or metabolic disease and/or two or more major coronary risk factors), or "known disease (cardiac, pulmonary, or metabolic)." The ACSM recommends that prior to vigorous exercise, exercise testing be conducted for apparently healthy men over age 40, apparently healthy women over age 50, persons with two or more cardiac risk factors, or persons with known disease.[26]

Routine, screening, resting ECGs have minimal predictive value in determining the health risk to a healthy respirator worker during periods of exertion. Several resting ECG findings (ST-segment depression, T-wave inversion, Q waves, and left-axis deviation) are associated with an increased risk of a coronary event.[27] These findings are uncommon in asymptomatic individuals but are not specific for CAD. As the pretest probability of CAD increases, so does the value of the ECG. Physicians or LHCPs may order a resting ECG to satisfy regulatory criteria or when clinically indicated. LHCPs may initiate the ECG, but a physician or midlevel clinician must interpret it and determine the need for further follow-up.

The exercise stress test (EST) more accurately predicts future coronary events than the resting ECG. The EST may be valuable in the further assessment of resting ECG abnormalities.[27] However, only 1 to 12 percent of asymptomatic persons with a positive exercise ECG suffer an acute myocardial infarction or sudden death.[24,28] Moreover, most serious cardiac events will occur in people with a negative EST. Given their cost and problems with sensitivity and specificity, ESTs must not be performed indiscriminately. Only a physician or midlevel clinician can prescribe an EST. In making this decision, the clinician should consider the work environment, work effort, respirator type, and cardiac risk factors (Table 3-7).

The EST also helps to characterize the work load that an individual with known cardiac disease can tolerate. The work output of a person performing an EST can be measured in metabolic equivalents (METs). METs provide a convenient scale for comparing performance on an EST with the cardiovascular demands of labor (see Table 2-3). An individual who expends 10 METs during a stress test without ECG abnormalities is unlikely to have clinically significant CAD. Similarly, this individual should have the functional capacity to perform most work activities with a respiratory protective device, PPE, and heat stress. Performing the test with the selected respirator and PPE provides even more assurance of the individual's functional capacity.

Table 3-7. Cardiac Risk Factors

1. Men > 45 years and women > 55 years or with premature menopause without estrogen replacement
2. Family history of myocardial infarction or sudden death before age 55 for any first-degree male family member or before age 65 for any first-degree female family member
3. Current cigarette smoker
4. Blood pressure greater than ≥ 140/90 mmHg, confirmed on two separate occasions
5. Hypercholesterolemia, with total cholesterol > 200 mg/dl or high-density lipoprotein (HDL) cholesterol < 35 mg/dl[a]
6. Diabetes mellitus
7. Sedentary lifestyle

[a]High serum HDL cholesterol (> 60 mg/dl) is considered a negative risk factor, i.e., if HDL is high, subtract one risk factor.

Source: Adapted from Summary of the Second Report of the National Cholesterol Education Program Expert Panel on Detection, Evaluation and Treatment of High Blood Pressure and Cholesterol in Adults. *JAMA* 1993; 269:3015–3023.

Olfactory Testing

Despite OSHA's new mandate for end-of-service-life indicators (ESLI) and change-out schedules, olfaction remains an important sense that warns respirator users of poor facial seal, exhaustion of the purification medium, other causes of respirator failure, or even the need to don a respirator. Therefore, people who report a diminished sense of smell should be further evaluated. Within their scope of practice, LHCPs may choose from several clinical tools to assess olfactory acuity. Abnormal tests should prompt consultation with a physician or midlevel clinician.

One of the OSHA qualitative fit-test methods, the isoamyl acetate protocol, begins with odor threshold screening to determine if the individual tested can detect the odor of isoamyl acetate at low levels. Failure to detect this odor requires the use of a different fit-test method but does not necessarily mean that the individual has a significant olfactory disorder.

The most popular clinical method to test olfaction resembles the OSHA protocol. The clinician presents four simple odorants to a subject and asks whether an odor is perceived and, if so, whether it can be identified.[29] Unfortunately, this method lacks sensitivity to many olfactory problems. In addition, even normal subjects may not be able to characterize the odors.[30] One simple improvement to this approach is the addition of a control vial. Appendix 6 outlines an olfactory screening procedure.

Commercial smell identification tests are also available. One example of a self-administered test method, the Pocket Smell Test, contains a booklet with four "scratch and sniff odorants" and an answer key. The subject must choose one of four multiple-choice answers for each odorant. This "forced choice" method is highly reliable and obviates the need for controls.[31]

Workers unable to correctly identify all four odorants should proceed to a more comprehensive screening evaluation. We use Sensonics' Smell Identification Test, which consists of 40 odorants in four booklets. The test, validated in a number of studies, compares responses to normative data by age and sex.[30]

In some cases, a subject may be hyposmic but still have sufficient olfactory capacity to detect relevant hazardous substances. Accordingly, PLHCPs may want to assess a worker's capacity to smell a specific substance in the workplace. Ideally, this test should be conducted with a control, as discussed earlier. Moreover, the examiner should take care not to expose the subject to hazardous concentrations of an agent. Approximate odor thresholds have been identified for many agents and can be compared with appropriate exposure limits.[32] Olfactory testing of specific agents should be conducted only when the odor thresholds are substantially lower than exposure limit concentrations.

Work Trial

The ordinary tools of a medical evaluation cannot always provide sufficient information for a PLHCP to make a rational recommendation regarding respirator use. In such cases, a work trial (WT) in a controlled environment can provide both the PLHCP and the worker with additional objective data about worker performance and respirator tolerance. PLHCPs can use the WT to assess work capacity and respirator tolerance in individuals with pulmonary conditions, claustrophobia, anxiety, sensory disturbances, and a variety of other health problems. WTs also can serve a valuable role in identifying effective accommodations for employees who need them. On the other hand, WTs cannot provide useful information for people with episodic conditions such as epilepsy or hypoglycemia. In addition, WTs are contraindicated if a worker has a known diagnosis or symptoms that could result in a catastrophic event precipitated by the demands of the trial. A work trial should *not* be confused with a cardiac EST. Although a cardiologist may use ambulatory ECG monitoring to assess rhythm abnormalities or silent ischemia, a WT should never be used for identifying cardiac abnormalities.

The respirator literature frequently recommends WTs as a means of assessing worker respirator tolerance. Unfortunately, the same literature provides little information about how to actually conduct such a trial, let alone assess the worker's performance.[33] We have devised WT protocols drawing on ACSM guidelines for exercise testing.[26] PLHCPs should understand the distinction however between ACSM-endorsed exercise testing and WTs. ACSM guidelines assume a gradual increase in energy expenditure and do not take into consideration the additional physiologic cost of the respirator or additional PPE. In a WT, increases in an employee's work load may not occur in a stepwise manner, the trial does not follow standardized exercise protocols, and the trial is conducted with the worker wearing all relevant PPE. These distinctions emphasize the importance of clearing an individual medically for a WT before it begins. The WT cannot reproduce all possible work situations. Most obviously, true emergencies and the resulting psychophysiologic response to them cannot be replicated. PLHCPs must use clinical judgment on a case-by-case basis when assessing worker tolerance. Because of the potential adverse health outcomes of a WT, LHCPs should rely on a physician or midlevel clinician to order trials that evaluate psychophysiologic limitations. Physicians and LHCPs should design WT procedures and assessment parameters with appropriate representatives of the employer.

Appendix 7 provides detailed procedures for two approaches to assessing respirator tolerance with a WT. In all cases, the worker must be medically evaluated first and be fit-tested prior to participation in the trial. WT protocols always must include appropriate work and medical safety procedures. WTs should never be conducted in immediately dangerous to life or health (IDLH) environments.

The level 1 WT addresses safety issues that may arise as the result of sensory or communication deficiencies. Trials may include evaluation of the impact of diminished hearing, smell, or vision. PLHCPs should recommend that this type of WT take place at the actual work site under normal working conditions, with attention to procedures to help ensure the employee's safety (see Appendix 7). Certain measures described below help ensure the safety of the employee. A nonclinical person with knowledge of the job tasks, such as a supervisor, may conduct the trial and document the results.

The level 2 WT explores potential psychophysiologic limitations of the worker and usually requires clinical expertise during the monitoring process. These WTs are designed to assess the significance of claustrophobia, panic attacks, altered respiratory function, or health conditions that may affect the overall fitness required for the job. These trials are conducted most often in a simulated work environment to avoid the potential adverse effects of respirator intolerance or failure. The employee wears all PPE required and performs the mechanics of the job. Examples of simulated environments include

- The actual work environment but without hazardous exposures
- A work area outside the actual workplace but in which the worker performs similar tasks
- A clinical area where makeshift tools and activities simulate the work, in particular the energy requirements.

The trial protocol includes all expected work activities, performed at the usual pace, such as climbing ladders, using tools, and handling materials. Indications for stopping a WT prior to its conclusion are listed in Appendix 7.

Before ordering a level 2 WT, the PLHCP must first, with the employer's help, identify the intensity of work expected for completing job tasks (see Table 2-3). The METs table does not take into account the physiologic impact of respirator use.

The clinical monitoring conducted during the 30-minute WT includes both subjective and objective data. The objective data involve measuring blood pressure, pulse, and respiration. Physical signs such as skin pallor, diaphoresis, disequilibrium, and respiratory distress are also documented. Important subjective information includes employee reports of anxiety, claustrophobia, light-headedness, and fatigue. A particularly useful tool is Borg's CR-10 scale (rating of perceived exertion). The scale, often used in exercise testing, correlates with both increases in oxygen uptake (Vo_2) and heart rate.[34] The CR-10 scale is sensitive to the differences in individuals' perceptions of the physical exertion required to complete a task (Table 3-8). As a general rule, few individuals are capable of sustaining near-maximal output for long, whether measured objectively or subjectively. Current recommendations for a sustainable work load over an 8-hour shift suggest a limit of 30 to 40 percent of the individual's Vo_{2max}.[35]

Heart rate is an important indicator during the clinical monitoring process because it correlates directly with exercise tolerance. Heart rates in healthy individuals participating in graded exercise plateau just prior to reaching Vo_{2max}.[26] A worker's maximal heart rate (±15 beats per minute) can be estimated by the formula (220 – age in years). Jackson[36] has developed a model that uses self-reported activity, sex, age, and body composition to more closely predict the functional aerobic capacity of an individual without the need to rely on exercise testing. These considerations provide a basis for assessing an individual's conditioning and for terminating a WT to avoid adverse health effects. The ACSM recommends that poorly conditioned individuals not exceed 60 percent of their maximal heart rate when beginning an exercise program.[37] Trained athletes may safely sustain up to 85 percent of their maximal predicted heart rate. For purposes of a WT of no longer than 30 minutes, we recommend 60 percent of a deconditioned person's maximal predicted heart rate as a point for test termination.[26] Conditioned, healthy workers may be allowed to exercise to 85 percent of their maximal predicted heart rate. Workers on beta blockers may only obtain a peak heart rate of 50 to 60 percent of the maximal heart rate predicted for their age. Additional criteria, as described in Appendix 7, must be used to assess these workers' WT tolerance.

Successful completion of a WT involves several criteria. First, depending on the indications for the trial, the employee must

Table 3-8. Borg's CR-10 Perceived Exertion Scale

0	Nothing at all
0.5	Extremely weak (just noticeable)
1	Very weak
2	Weak (light)
3	Moderate
4	
5	Strong (heavy)
6	
7	Very strong
8	
9	
10	Extremely strong (almost max)
•	Maximal

successfully perform the required job tasks, communicate effectively, or identify relevant odors. Second, the employee must complete the WT without exceeding any of the health criteria for stopping the exercise. Third, the PLHCP must review the results of the WT and determine that the individual's performance would be sustainable for the actual time period and conditions required by the employer.

If an individual does not complete a WT successfully, the trial may be repeated with an accommodation appropriate for their medical condition (see Chapter 4). If the employee meets the criteria for a successful WT with accommodations, the PLHCP can approve the individual for respirator use with specific accommodations.

Conclusion

Medical evaluation for respirator use can be as simple as reviewing a questionnaire or as complex as performing detailed medical examinations involving consultants, advanced laboratory investigation, and work trials. This chapter has reviewed procedures for the administration of a variety of tools useful in the medical evaluation. The next chapter provides guidelines for interpretation of the data collected with the these tools.

Further Reading

OSHA Directives: CPL 2-0.120—Inspection Procedures for the Respiratory Inspection Standard. Available at *http://www.osha-slc.gov/Publications/CPL_2-0_120.pdf. This directive provides an overview of procedures for an OSHA inspection of a company's respiratory protection program.*

American College of Occupational and Environmental Medicine. ACOEM Position Paper. Spirometry in the Occupational Setting. Available at *http://www.acoem.org/paprguid/papers/sppaper/spirom.htm. An updated review of the various uses of spirometry in the occupational setting along with all the information necessary to conduct valid and reliable tests. Clinicians will find the information about spirometric interpretation particularly useful.*

American Thoracic Society. Lung function testing: Selection of reference values and interpretative strategies. *Am Rev Respir Dis* 1991; 144:1202–1218. *Classic reference for selection of spirometry reference values and interpretation of spirometry test results.*

American College of Sports Medicine. *ACMS's Guidelines for Exercise Testing and Prescription*, 5th ed. Baltimore: Williams & Wilkins, 1995. *Good guideline for physical fitness testing that provides a framework for the evaluation of a worker's physical condition in relation to job demands.*

Borg G. Psychophysical scaling with applications in physical work and the perception of exertion. *Scand J Work Environ Health* 1990; 16(suppl 1):55–58. *This article provides details on the use of the Borg ratings of the perceived exertion scale.*

Doty RL, Kobal G. Current trends in the measurement of olfactory function. In Doty RL (ed.), Handbook of Olfaction and Gustation. New York: Marcel-Dekker, 1995:191–225. This chapter explains all that most clinicians will ever need to know about assessing olfactory functioning. Interested readers will be intrigued by other chapters of this book, which is rich in details about the wonders of the physiology of taste and smell.

References

1. Respiratory Protection Preamble: VI. Summary of the Final Economic Analysis. Available at *http://www.osha-slc.gov/Preamble/RP_html/RESPIRATORY6.html#TEXT1.*
2. Small Entity Compliance Guide for Revised Respiratory Protection Standard. Available at *http://www.osha-slc.gov/Publications/secgrev-current.pdf.*
3. OSHA Standards: Compliance and Interpretation Letters. Available at *http://www.osha-slc.gov/OshDoc/Interp_data/I19981016A.html.*

4. Harber P, Merz B, Chi K. Decision model for optimizing respirator protection. *J Occup Environ Med* 1999; 41:356–365.
5. Harber P. Respirators. In Rom WN (ed.), *Environmental and Occupational Medicine*, 3d ed. Philadelphia: Lippincott-Raven, 1998:1763–1764.
6. Pappas GP, Takaro TK, Stover B, et al. Rates of medical clearance and causes for work restrictions. *Am J Ind Med* 1999; 35:390–394.
7. Respiratory Protection Preamble: VII. Summary and Explanation. Available at *http://www.osha-slc.gov/Preamble/RP_html/RESPIRATORY7.html.*
8. Pappas GP, Takaro TK, Stover B, et al. Medical clearance for respirator use: Sensitivity and specificity of a questionnaire. *Am J Ind Med* 1999; 35:395–400.
9. American Thoracic Society. Respiratory protection guidelines. *Am J Respir Crit Care Med* 1996; 154:1153–1165.
10. 10 CFR Part 20.1073. Available at *http://www.nrc.gov/NRC/CFR/PART020/part020-1703.html.*
11. Muhm JM. Medical surveillance for respirator users. *J Occup Environ Med* 1999; 41:989–994.
12. American National Standards Committee on Respiratory Protection. Z88: American National Standard for Respiratory Protection—Respirator Use: Physical Qualifications for Personnel (ANSI Z88.6-1984). New York: American National Standards Institute, 1984.
13. *NIOSH Respirator Decision Logic.* DHHS (NIOSH) Publication No. 87-108. Washington: U.S. Department of Health and Human Services, Public Health Service, Centers for Disease Control, National Institute for Occupational Safety and Health, Division of Safety Research, 1987.
14. Lerner PJ (ed). Occupational medicine forum: What is the recommended frequency and content of respirator user medical evaluations? *J Occup Environ Med* 1998; 40:109–111.
15. American Thoracic Society. Lung function testing: Selection of reference values and interpretative strategies. *Am Rev Respir Dis* 1991; 144:1202–1218.
16. Boehlecke BA. Respirators. In Harber P, Schenker MB, Balmes JR (eds.), *Occupational and Environmental Respiratory Disease*. St. Louis: Mosby–Year Book, 1996:963–971.
17. American Thoracic Society. Standardization of spirometry, 1994 update. *Am J Respir Crit Care Med* 1995; 152:1107–1136.
18. Pulmonary Function Standards for Cotton Dust Standard—29 CFR 1910.1043, Appendix D. Available at *http://www.osha-slc.gov/OshStd_data/1910_1043_APP_D.html.*

19. Hankinson JL, Odencrantz JR, Fedan KB. Spirometric reference values from a sample of the general U.S. population. *Am J Respir Crit Care Med* 1999; 159:179–187.
20. ACOEM Position Paper. Spirometry in the Occupational Setting. Available at *http://www.acoem.org/paprguid/papers/sppaper/spirom.htm.*
21. ACOEM Position Paper. Spirometry in the Occupational Setting: Spirometry Interpretation Flow Chart. Available at *http://www.acoem.org/paprguid/papers/sppaper/table6.5.htm.*
22. Knudson RJ, Lebowitz MD, Holberg CJ, et al. Changes in the normal maximal expiratory flow-volume curve with growth and aging. *Am Rev Respir Dis* 1983; 127:725–734.
23. Conner B. Comprehensive care in the allergy/asthma office: Pulmonary function testing. *Immunol Allergy Clin North Am* 1999; 19:48–61.
24. U.S. Preventive Services Task Force. *Guidelines from Guide to Clinical Preventive Services,* 2d ed. Baltimore: Williams & Wilkins, 1996.
25. American Academy of Family Physicians. *Age Charts for Periodic Health Examination.* Reprint No. 510. Kansas City, Missouri: American Academy of Family Physicians, 1994.
26. American College of Sports Medicine. *ACSM's Guidelines for Exercise Testing and Prescription,* 5th ed. Baltimore: Williams & Wilkins, 1995.
27. Kwok JM, Miller TD, Christian TF, et al. Prognostic value of a treadmill exercise score in symptomatic patients with nonspecific ST-T abnormalities on resting ECG. *JAMA* 1999; 282:1047–1053.
28. Chaitman B. Exercise stress testing. In Braunwald E (ed.), *Heart Disease: A Textbook of Cardiovascular Medicine,* 5th ed. Philadelphia: Saunders, 1997:153–174.
29. Bates B. *A Guide to Physical Examination,* 3d ed. Philadelphia: Lippincott, 1983.
30. Doty RL, Kobal G. Current trends in the measurement of olfactory function. In Doty RL (ed.), *Handbook of Olfaction and Gustation.* New York: Marcel-Dekker, 1995:191–225.
31. Doty R, Monroe C. Occupational screening of olfactory function. In *Occupational & Environmenta lChemical Hazards.* Chichester: Ellis Horwood, 1987.
32. American Industrial Hygiene Association. *Odor Thresholds for Chemicals with Established Occupational Health Standards.* Fairfax, Virginia: American Industrial Hygiene Association, 1989.
33. Muhm JM. Medical surveillance for respirator users. *J Occup Environ Med* 1999; 41:989–994.
34. Borg GA. Psychophysical scaling with applications in physical work and the perception of exertion. *Scand J Work Environ Health* 1990; 6(supp 1):55–58.

35. NIOSH Criteria Documents. Criteria for a Recommended Standard Occupational Exposure to Hot Environments. Revised Criteria 1986. DHHS (NIOSH) Publication No. 86-113. Washington: National Institute of Occupational Safety and Health, 1986.
36. Jackson AS. Prediction of functional aerobic capacity without exercise testing. *Med Sci Sports Exerc* 1990; 22:863–870.
37. Sudy M (ed). *Personal Trainer Manual: The Resource for Fitness Instructors.* San Diego: American Council on Exercise, 1991.

4

Medical Decision Making for Respirator Approval

After collecting the initial set of medical data by questionnaire or examination, the physician or licensed health care professional (PLHCP) must decide whether to approve the examinee for respirator use. This chapter provides guidance for this decision making. Before presenting the guidelines, however, introductory sections consider (1) the overarching questions that must be answered during the medical evaluation, (2) the predictive value of the medical assessment, and (3) the relevance of the Americans with Disabilities Act.

The Overarching Questions

To make a decision about qualifying an individual for respirator use, the physician or LHCP must answer three overarching questions:

1. Can the individual safely wear a specific type of respirator while performing a particular task in the work environment?
2. Will the worker comply with respirator use?
3. What may be the consequences of respirator failure?

Question 1 involves considering the interactions of the individual, the respirator, and the work. Question 2 forces the examiner to consider whether the individual will tolerate the discomfort and inconvenience of respirator use. Respirator compliance and efficacy in preventing disease depend not only on the individual's health and comfort but also on the company's culture and the comprehensiveness of its respiratory protection program.[1] Question 3 requires that physicians or LHCPs incorporate a judgment about the seriousness of respirator failure in their decision making. Respirator noncompliance or intolerance resulting in compromised work performance may matter little when the hazard is nuisance dust but may be life-threatening in an immediately dangerous to life or health (IDLH) atmosphere.

The medical evaluation may identify three types of reasons for limiting an individual's use of a respirator: psychophysiologic impairment, episodic incapacitation, or working conditions. Workers' psychophysiologic conditions may limit their respirator and exertion tolerance. For example, coronary artery disease may impair an individual's ability to exercise beyond a certain level of exertion. Alternatively, people may have conditions that normally allow them to perform their job duties while wearing a respirator but episodically disable them. A person with active epilepsy or asthma normally can perform all job tasks with a respirator but could be put at grave danger if incapacitated in an IDLH atmosphere. Working conditions and exposures may significantly aggravate an individuals' underlying medical problems. A worker with ischemic cardiac disease may be able to wear a respirator in temperate but not hot environments. An individual sensitive to toluene diisocyanate may die if respirator failure occurs while working with this agent.

Predictive Value of the Medical Evaluation

The decisions of physicians and LHCPs about respirator approval should not be arbitrary or based solely on opinion. They should be based on objective data about the examinee's health, the work requirements, and the respirator, with reference to the relevant medical literature. In an ideal world, sufficient scientific information would be available to predict with 100 percent accuracy whether a worker could perform a job safely while wearing a respirator. Reality disappoints. Other than past respirator tolerance, research has not identified reliable medical predictors of successful respirator use.[2]

Harber et al.[3] describe a shift over time in the medical basis for respirator certification. Considerations have included symptoms, diagnosis, clinical judgment, abnormal laboratory tests such as spirometry, and physiologic limitations demonstrated by an inability to perform a simulated work task. Traditional practice in restricting respirator use typically depends heavily on individual clinician idiosyncrasies and convention, codified in guidelines such as ANSI Z88.6-1984,[4] NFPA 1582,[5] and the Nuclear Energy Institute (NEI) Personnel Access Data System (PADS) Project.[6] Other safety-sensitive industry guidelines such as those from the Nuclear Regulatory Commission (NRC),[7] the Department of Transportation (DOT),[8] and the Federal Aviation Administration (FAA)[9] also have relevance because they contain recommendations about medical conditions that may cause sudden incapacitation or chronic impairment. Increasingly, as these guidelines have been updated, they have avoided absolute proscriptive medical standards, ultimately leaving qualification up to the clinical judgment of the examiner.

In many cases, these guidelines have face validity. For example, given that it is known that respirators place additional cardiopulmonary stresses on users, abnormal spirometry has been used as grounds for denying respirator approval. Similarly, a person with recurrent severe hypoglycemic episodes, who therefore may be incapable of safely leaving a hazardous environment, may not be a good candidate for a self-contained breathing apparatus (SCBA).

Improper use of respirators has caused significant adverse health effects.[10] However, despite theoretical concerns, respirators themselves rarely have been documented to precipitate serious health problems.[11] Possibly, this paucity of respirator-related adverse health events in the literature results from poor reporting and investigation. Nonetheless, the evidence suggests that in most cases, if a person is otherwise medically qualified to do a job, that individual should be able to tolerate the additional stresses of a respirator.[2] In every case, decision making therefore must include consideration of whether the individual is capable of performing the work that requires a respirator. The job tasks themselves may be more physically stressful than the physical demands associated with respirator use.[12]

When available, relevant scientific literature is the criterion we cite as the rationale for decision-making recommendations. Since community standards often are applied in evaluating the quality of medical decision making, relevant published guidelines commonly used in the occupational medicine community are also referenced. As with other preplacement examinations, the ideal predictor of successful use is a history of *past* successful respirator use under similar work conditions. In borderline cases, we liberally recommend conducting work trials incorporating an informed employee's judgment as one of the best arbiters. Without sound medical or performance-based evidence, physicians and LHCPs should avoid denying or restricting respirator use unless employees express a desire to avoid the attendant risks. Because of the difficulties in predicting respirator tolerance, managers of respiratory protection programs should place more emphasis on medical reassessment of individuals after they have begun wearing respirators. The guidelines in this chapter provide condition-specific recommendations for reevaluation.

Americans with Disabilities Act

Medical decision making assumes importance that goes beyond the simple question of respirator approval. Denial or restriction of respirator use may threaten a worker's job or compensation. Since 1992, the Americans with Disabilities Act (ADA) makes the employer, and possibly the physician or LHCP, liable for discriminatory employment decisions that affect the worker. Practitioners also should consult additional state or country statutes that may pertain in some situations.

Employers, not physicians or LHCPs, make final decisions regarding respirator use, employment, and reasonable accommodations. Nonetheless, employers rely heavily on medical guidance. Therefore, before making recommendations about respirator use to an employer, physicians or LHCPs must consider whether their decisions conform to the standards of the ADA. Chapter 2 notes that the ADA, as well as Occupational Safety and Health Administration (OSHA), requires the confidentiality of employment-related medical examinations. This section briefly presents additional information that physicians and LHCPs should consider in advising employers about an individual's respirator use. Readers needing more background about the ADA should consult the Further Reading section at the end of the chapter.

Fundamentally, the ADA prohibits employment discrimination against qualified people with disabilities. People are protected by the ADA if they have a disability, had a disability, or are perceived to have a disability. The ADA defines a disability as a physical or mental impairment that substantially limits one or more major life functions, including breathing, walking, standing, performing activities of daily living, and working. Recent Supreme Court decisions have constrained the concept of impairment to conditions that cannot be medically corrected. The Court ruled that disorders such as correctable visual problems or hypertension cannot be claimed as disabilities covered by the ADA.

Employers may use information provided by physicians or LHCPs in making employment decisions as long as this information is relevant to the individual's work tasks. However, the ADA prohibits discriminatory employment decisions made if a person's disabilities interfere only with nonessential aspects of a job. If essential job duties do not require respirator use, employers cannot exclude individuals from work simply because they are unable to wear a respirator.

The ADA does not require an employer to hire or retain an individual unable to use a respirator for an essential task if such use would pose a direct threat to the health and safety of the individual, coworkers, or the public. However, the ADA establishes a high hurdle to meet the standard of direct threat. Four considerations are used to assess this threat: the nature and severity of the potential harm, the likelihood of harm, the imminence of the threat, and the duration of risk. *Severity* refers to the consequence of the interaction between work and an individual's medical condition. A myocardial infarction meets this criterion better than a skin rash. *Likelihood* refers to the probability that an adverse effect will occur. Speculation and unsupported "opinion" do not meet this test. To attain the standard of *imminence,* the adverse consequence must be expected to occur in the very near future. A problem that may occur in the remote future cannot form the basis of exclusion. Finally, *duration of risk* refers to whether the condition is permanent or instead may be treatable or reversible.

Few respirator examinations are likely to identify conditions that meet the ADA criteria for a direct threat that cannot be minimized by appropriate accommodations. In fact, at least one study concludes that few workers fail medical clearance for respirator use or receive work restrictions.[13] Of 5569 Department of Energy workers reviewed, only 0.2 percent were denied medical clearance. Another 1.1 percent received work restrictions. Pregnancy was the most common cause for denying medical clearance for respirator use. Lung disease, cardiovascular disease, and claustrophobia represented the most common causes for work restrictions. Physical examination and spirometry added little to the detection of conditions that led to restrictions.

If the physician or LHCP concludes that respirator use does present a direct threat to the individual, coworkers, or the public, the employee may request an accommodation by the employer. In some cases, the physician or LHCP may help identify a reasonable accommodation that will allow the worker to accomplish the essential functions of a job without such concerns. The ADA does not specifically define *reasonable* but notes that the critical consideration concerns the hardship such an accommodation places on the employer. Larger employers with greater resources are expected to make greater accommodations. Accommodations may involve a change in the physical demands of a job, modification of working conditions, administrative controls, or individual interventions. Mohm[14] has recommended specific accommodations related to specific symptoms and respirator types. Correlating specific respirator components with their physiologic burden, he has made detailed recommendations regarding adjustments that can be made to a respirator for specific symptoms and medical conditions (see Table 4-1). The guidelines provide other examples of accommodations.

Table 4-1. Recommended Approaches to Modifying Respirators for Symptomatic Users

Symptom	Medical conditions	Respirator type	Recommended accommodations
Dyspnea	Obstructive lung disease Restrictive lung disease Neuromuscular disease Anxiety disorder	Air-purifying	Supply (PAPR)[a] Check function of air flow valves Replace canister/ cartridge/filter Reduce mask size Add airflow baffles
	Neuromuscular disease Anxiety disorder	Air-supply	Adjust regulator
	Coronary artery disease		
Angina/arrhythmia/CHF[a]	Conduction system disease Myocardial dysfunction	Air-purifying	Supply PAPR Check function of airflow valves Replace canister/ cartridge/filter Reduce mask size Add airflow baffles
		Air-supply	Adjust regulator
Claustrophobia	Anxiety disorder	Either	Loose-fitting helmet
Fatigue/mental clouding	Hyperthermia	Either	Adjust work/rest cycles Cooling vests Supply PAPR
Skin irritation	Pseudofolliculitis barbae Dermatitis Latex allergy	Either	Loose-fitting helmet or for latex allergy, latex-free, silicone mask

[a]PAPR = powered air purifying respirator; CHF = congestive heart failure.
Source: Adapted from Muhm JM. Medical surveillance for respirator users. *J Occup Environ Med* 1999; 41:989–994.

Using Medical Data to Make Decisions About Respirator Approval

The following guidelines provide comprehensive protocols for using medical data to evaluate an individual's ability to wear a respirator safely. In many cases, PLHCPs may appropriately make these decisions using only responses to the questionnaire. In other situations, they will need to obtain more information by interview or from the treating physician(s). In a smaller number of circumstances, high-quality decision making demands referral to practitioners with greater clinical and occupational expertise and practice privileges. In each section of the protocol, guidance for respirator approval begins with the simplest scenarios. The protocol then progresses in complexity to cases that *require* consultation with a clinician with more advanced knowledge and experience relevant to the clinical scenario. Usually a nurse-practitioner, physician's assistant, or physician with specialized knowledge of occupational health and respirators conducts the more advanced medical evaluation. If an advanced clinician with occupational medical expertise is involved from the beginning of the evaluation, referral to another practitioner or physician seldom will be necessary. In some cases, however, PLHCPs should refer to physicians with other relevant medical expertise to better define an individual's health problem and optimize its treatment. PLCHPs should not, however, abdicate their role in respirator qualification. Rather they should incorporate the opinions of consulting physicians into the decision-making process. State regulations and individual hospital privileges must be consulted to ascertain the PLHCP's scope of practice.

The organization of the guidelines parallels the OSHA standard by presenting the physician or LHCP with two approaches for finding the relevant protocol. Readers with questions about decision making related to specific organ system disorders identified by a traditional history and physical examination may look for guidance by finding the appropriate organ system. Each major heading of the guidelines also corresponds to questions from the OSHA questionnaire. Readers can identify the relevant protocol by looking for the highlighted question drawn from the OSHA questionnaire (Appendix 1C). Because the questionnaire is redundant, multiple related questions are grouped under a single organ system heading.

The protocols also address some common disorders relevant to respirator use that may be identified during a medical examination, although probably not by the questionnaire. This review does not attempt, however, to second as a textbook of medicine. Undoubtedly, physicians and LHCPs will encounter other significant health problems that could affect respirator use. In most cases, irrespective of respirator use, these problems would substantially impair an individual from performing the job tasks themselves.

Each section of the guidelines begins with a general discussion of aspects of the identified health problem relevant to respirator use. The next segment ("Interview") outlines an approach to gathering more information during a follow-up evaluation. This approach relies on an oral or written interview of the employee and treating clinician. Recommended generic questions for this interview are detailed in Figure 4-1. Obtaining medical records may substantially delay the decision-making process. PLHCPs should use their judgment in deciding whether the documentation provided by treating physician records is crucial to the assessment. Faxing specific questions to attending physicians with prewritten, check-off answers expedites the turnaround of information. The third segment ("Assessment") provides guidance about analysis of this additional information. The final segment ("Plan") recommends a course of action based on the assessment. This segment indicates when advanced medical evaluation is appropriate. The plan also addresses worker education and potential accommodations for overcoming health problems that may interfere with safe respirator use.

In the absence of any positive responses to questions 1 to 8 and 10 to 15 of the OSHA-mandated questionnaire, PLHCP can approve the examinee for respirator use without restrictions. We suggest prompting periodic reevaluations as discussed in Chapter 3 in addition to the event-based reevaluations prescribed by OSHA. The respiratory protection standard states that a positive response to any of the mandatory questions (Part A, Section 2, questions 1 to 15, with the exception of question 9) requires a follow-up medical evaluation, as described previously. Although not clearly stated in the standard, OSHA clarified in a later publication that answering yes to questions 10 to 15 also requires medical follow-up.[15] Questions 10 to 15 become mandatory only if the worker will be wearing a full-mask respirator or an SCBA. The guidelines prescribe appropriate follow-up for questions 1 to 15 or similar data collected from a traditional medical evaluation. To obviate the need for referral for further medical evaluation, PLHCPs often can gather sufficient additional information through interview of the examinee and written information from the primary care physician. Figure 4-1 provides a template for requesting the additional information specific to each question.

Figure 4-1. Respirator Use Medical Evaluation Supplemental History

Name: ______________________ DOB __________ Date ________

Diagnosis: __________________________ Date of Onset: ________

Etiology: ____

Current symptoms: [ ] no [ ] yes; describe: ________________

Episodic symptoms: [ ] no [ ] yes; describe: ________________

Warning present before incapacitation [ ] no [ ] yes

Describe: ______________________________________

Number of episodes in last year: ____________________

Date of most recent episode: _______________________

Triggers: ____

Diagnostic tests: ____________ ____________ ____________

date test result

____________ ____________ ____________

date test result

Medication and dosage:

Other treatment: ____

Effectiveness of treatment: [ ] good [ ] poor

Explain: ____

Hospitalizations: [ ] no [ ] yes: [ ] Procedures: ______________

____

Adequate knowledge about disease and its management:

[ ] no [ ] yes; explain: _____________________________________

Past problems with respirator use due to this condition:

[ ] no [ ] yes; explain: _____________________________________

Date last respirator use: __________ Respirator type: ______________

Treating clinician: ______________________ phone # ____________

PLHCP signature ______________________ Date ______________

We believe that denial of respirator use, as well as clinical investigations and treatment that fall outside the LHCP's scope of practice, should be referred to the appropriate advanced clinician with expertise in occupational medicine and respirator evaluation. If the worker refuses referral, the physician or LHCP may recommend that the employer deny or appropriately restrict respirator use "pending further medical evaluation."

Since the predictive value of medical data for safe respirator use remains uncertain in many cases, we rarely recommend categorical denial of respirator use because of information obtained during the initial medical examination. Instead, every attempt should be made during follow-up evaluation to obtain data that can improve the clinician's assessment.

Medical Guidelines for Respiratortor Use

Tobacco Use

1. Do you currently smoke tobacco, or have you smoked in the last month?

Tobacco use can impair the respiratory and cardiovascular systems sufficiently to affect an individual's ability to perform work safely while wearing a respirator. Further information about the extent of tobacco use, as well as about clinical symptoms, may be collected through personal interview. Employers interested in smoking cessation may agree to an age-based schedule of periodic additional examinations that allows for frequent repeated recommendation of smoking cessation as well as close attention to the relatively high risk of relevant end-organ damage that smokers suffer. The Agency for Health Care Policy and Research (AHCPR) clinical practice guideline recommends that clinicians regularly discuss smoking cessation with smokers.[16]

A report of tobacco use in the absence of complicating medical problems should not precipitate respirator disqualification or restrictions. On further interview, however, a worker may admit to medical problems not reported initially. Follow-up with these individuals also presents the opportunity to reinforce the importance of smoking cessation and personal hygiene with respect to the use of tobacco products. Employees need to understand the special risks associated with contamination of tobacco products with workplace contaminants, the need for hand washing before smoking, and the additive or synergistic toxicity of tobacco and many occupational hazards.

Interview

The PLHCP should interview the employee to gather additional information to further probe whether the worker's tobacco use has resulted in clinically apparent end-organ damage. As in any health care encounter, the PLHCP also should inquire about whether the worker is interested in smoking cessation.

Assessment

- Review and probe by interview answers to the respiratory, cardiovascular, and medication questions (3–7).
- Recommend clinical examination (respiratory, cardiovascular), including spirometry, if clinically significant positive responses are obtained to any of questions 3–7c (see protocols described below for each question answered affirmatively).
- Assess readiness for smoking cessation.

Plan

Respirator Approval Decision

1. Individual is asymptomatic, and there is no history of respiratory or cardiovascular disease or other limiting health conditions.
    - Approve for all respirators without additional consultation or evaluation. (A clinical examination is recommended prior to SCBA approval.)
2. The individual has any *chronic* or *recurrent* respiratory symptoms (question 4), active respiratory disease (question 3), cardiovascular symptoms (questions 5 and 6), or history of using cardiorespiratory drugs (question 7).
    - Defer respirator approval pending advanced medical evaluation.
    - Provide employee and employer with an appropriate timeline for completion of the evaluation.

3. Recommend reevaluation by brief questionnaire (see Table 3-3) or, if the employer concurs with the importance of repeated smoking cessation messages, by an age-related guideline (see Table 3-4).

Education

1. Hazards of tobacco use.
2. Additive and synergistic toxic effects of tobacco and occupational hazards.
3. Need for proper storage of tobacco products at work to avoid contamination.
4. Need for hand washing at work before smoking to avoid inhalation of contaminants.
5. Need to report change in health status to trigger medical reevaluation for respirator use.
6. Provide information about smoking cessation.
7. Report respirator intolerance.

Neurologic Conditions

2a. Have you ever had seizures (fits)?

A neurologic condition that results in episodic, unpredictable disturbances of awareness may be a reason for restricting respirator use. Loss of consciousness or psychomotor abilities while wearing a respirator in an IDLH atmosphere, while operating dangerous machinery, or while working at heights may result in death or serious health consequences. Incapacitation in less hazardous situations while wearing a respirator is much less likely to result in a significant problem. Generalized seizure disorder represents the most common neurologic condition that leads to episodic impairment of awareness. Although as many as 10 percent of people experience at least one seizure in their lifetime, less than 1 percent of the population has epilepsy—unprovoked recurrent seizures.[17] In other words, most people with a remote history of seizures are not likely to have another one. Narcolepsy and severe sleep apnea may cause irresistible urges to sleep during periods of tedium or while performing tasks that require sustained vigilance. Various disorders of the autonomic nervous system may result in neurogenic syncope. Cerebrovascular disease (see the section on cardiovascular disorders) may present as transient ischemic attacks and stroke with their attendant acute onset of neurologic impairment.

Many other neurologic conditions potentially may interfere with the safe use of a respirator. In general, the key consideration is whether the person can perform the job tasks irrespective of a respirator. To the extent that the medical condition may uniquely interfere with respirator use, the issue probably will revolve around a person's neuromuscular and cognitive ability to don and doff a respirator as required and have the strength, balance, and judgment to use it appropriately. Broadly characterized, the disorders of potential concern include disturbances of posture, balance, and gait, such as parkinsonism, multiple sclerosis, and neuromuscular diseases. Another group of potentially limiting diseases consists of those associated with extremity pain, numbness, and weakness, such as peripheral neuropathy.

Interview

The PLHCP should interview the employee to gather additional information about the seizure history (see Figure 4-1). In many cases, referral for further evaluation will not be necessary. Instead, after obtaining an appropriate release, the PLHCP should request the following information from the individual's attending clinician:

- Type of seizure disorder
- Etiology of the seizure disorder (e.g., trauma, idiopathic, metabolic, vascular, infection)
- Triggers
- Seizure control (average number of seizures per year and last known seizure)
- Aura or other warning signs that the individual may have a seizure
- Medication
- Persistent neurologic impairment or symptoms
- Primary care clinician's written opinion regarding respirator use with reference to the individual's specific job tasks

Assessment

Historically, employment discrimination has needlessly decreased opportunities of people with a history of well-controlled seizures. Many individuals with a history of epilepsy can perform most job tasks with a respirator safely. Little empirical information specific to respirator use is available.

As evidence of increasing confidence in the ability of people with controlled seizure disorders to operate dangerous machinery, most states now no longer require an absolute predefined seizure-free period for obtaining a driver's license. While states generally require a 6- to 12-month seizure-free period, many will allow exceptions under which a license may be issued after a shorter period. Possible exceptions include the following: a breakthrough seizure due to physician-directed medication change, an isolated seizure where the medical examination indicates that another episode appears unlikely, a seizure related to a temporary illness, a seizure due to an isolated incident of not taking medication, an established pattern of only nocturnal seizures, an established pattern of only seizures that do not impair driving ability, or an established pattern of an extended warning aura. In addition, some states make licenses available with restrictions such as daytime driving only, driving to and from work or within a certain distance from home, or driving only in an emergency.[18]

On the other hand, given the issues of public safety, the DOT continues to consider a history of seizures as being an absolute contraindication to holding a commercial driver's license. The DOT does make exceptions that are relevant to respirator approval.[19] Childhood febrile seizures or a remote history of seizures (more than 10 years ago) without ongoing use of medications need not result in disqualification. A history of a solitary seizure, without no recurrence off medications for 5 years, also does not result in ineligibility for a commercial driver's license.

NFPA 1582 designates epileptic conditions as disqualifying for firefighting unless the individual has had a provoked seizure with the precipitant identified and alleviated.[5] Further requirements include a subsequent normal electroencephalogram (EEG), normal neurologic examination, freedom from seizure recurrence for 1 year off medication, and a definitive statement from a qualified neurologic specialist.

In general, employees may be approved for respirator use with a seizure-free interval of greater than 6 to 12 months without other ongoing neurologic impairment or symptoms and the absence of medication side effects as long as the primary care clinician concurs. Exceptions to this seizure-free period may be made on the basis of the practice of state driver's licensing guidelines noted above. Recent seizures, changes in medication, psychomotor effects of medication, persistent neurologic impairment or symptoms, or concerns of the primary care clinician about an individual's ability to wear a respirator should prompt referral to an appropriate clinician for additional evaluation. The clinician should exercise more caution if the individual will wear a respirator in an IDLH atmosphere or public safety may be at significant risk.

Clinicians evaluating the capabilities of individuals with other potentially limiting neurologic conditions should rely heavily on information from the treating physician(s) along with observations made during the fit testing, functional capacity evaluations, or work trials to finalize opinions.

An affirmative answer to any of the following questions may indicate the need for respirator denial or restrictions and should prompt advanced medical evaluation.

- Has the employee had a seizure within the last 12 months?
- Has the employee changed or decreased his or her dose of medications within the last month?
- Does the employee take medications likely to interfere with safe job performance while wearing a respirator?
- Does the employee have a history of respirator intolerance or occupational problems due to seizures?
- Will the employee be working in conditions likely to trigger a seizure?
- Does the employee show a lack of understanding of the disease or compliance with medical recommendations?
- Will the employee be working in an IDLH atmosphere?

Plan

Respirator Approval Decision

1. Individuals with a history of early childhood seizure(s) secondary to fever/infection, an isolated seizure more than 5 years ago, or

multiple seizures more than 10 years ago without recurrence off medications may be approved for respirator use without restrictions barring any other medical concerns.

- Approve for all respirators without additional consultation or evaluation. (A clinical examination is recommended prior to SCBA approval.)
- Recommend reevaluation by brief questionnaire (see Table 3-3) or age-related guideline (see Table 3-4), barring any change in health status or working conditions.

2. Individuals with controlled seizure activity (no seizure on or off medications in the last year, medical regime stable, no medication side effects likely to impair work, and exposure to seizure triggers at work unlikely):
   - Approve without additional consultation or evaluation for all respirators to be used in a non-IDLH environment after receiving documentation from treating clinician that condition is stable and that there are no contraindications to respirator use. (A clinical examination is recommended prior to SCBA approval.)
   - If person will work in an IDLH environment, defer respirator approval pending advanced medical evaluation.
   - Provide employee and employer an appropriate timeline for completion of the evaluation.
   - Recommend follow up evaluation at a minimum of an annual basis or earlier if a change in treatment or health status occurs.
3. Individuals with poorly controlled seizure activity, changing medical regime, work exposures to triggers, or relevant side effects from medications:
   - Defer respirator approval pending advanced medical evaluation.
   - Provide employee and employer an appropriate timeline for completion of the evaluation.

Additional Considerations for Advanced Respirator Approval

Active seizure disorder (seizures less than 1 year ago on or off of medications) and other relevant neurologic conditions:

- Defer respirator approval pending advanced medical evaluation and optimization of therapeutic management.
- Provide employee and employer with an appropriate timeline for completion of the evaluation and treatment optimization.
- Consider obtaining a written recommendation from the attending or consulting neurologist.
- Consider the hazards of work the environment, the consequences of respirator failure, and whether recent seizures or other disturbances of awareness may be treatable or explained by controllable or predictable events.
- When appropriate, recommend accommodations and restrictions specific to the work site or recommend denial of respirator approval.

Education

1. Need for reevaluation for respirator use if health status, medications, or work conditions change.
2. Minimize any known triggers such as shift work, stress, and alcohol use.
3. Encourage employee to sign a release to allow notification of first-aiders about appropriate response if worker has a seizure at work.
4. Alert the physician or LHCP and employer of any need for accommodations.
5. Report respirator intolerance.

Potential Accommodations for Workers with Seizure Disorders

- Eliminate individual's seizure triggers, e.g., restrict shift work or exposure to flickering light.
- Allow temporary restrictions from all respirator use for planned changes in anticonvulsant regime.
- Eliminate job requirement to work in IDLH atmosphere.
- Eliminate job requirement to work with respirator.

The Epilepsy Foundation of America provides information about employment and seizure disorders[20] and also lists many ideas to help accommodate workers with seizure disorders.[21]

Endocrine and Metabolic Conditions

2b. Have you ever had diabetes (sugar disease)?

Diabetes, heat intolerance, and obesity lead the list of metabolic conditions that may affect respirator tolerance. The OSHA questionnaire only inquires directly about diabetes. Diabetes and obesity can lead to chronic complications considered in other sections of these guidelines. Heat intolerance may yield some nonspecific symptoms similar to those of other conditions identified in the questionnaire.

Diabetes

Episodic, unpredictable impairment of psychomotor abilities due to hypoglycemia presents the greatest endocrinologic concern for respirator use. The vast majority of people with type I diabetes experience many symptomatic episodes of hypoglycemia, probably between one and two per week. Annually, many persons with insulin-treated diabetes mellitus (ITDM) experience severe events, defined as requiring the assistance of someone else.[22] Research indicates that a minimum of 17 percent of people receiving conventional insulin treatment have at least one severe hypoglycemic episode per year.[23] Intensive insulin control triples this rate. Moderate hypoglycemia occurs much more frequently. Loss of consciousness, coordination, strength, fine motor skills, or cognitive functions can be fatal or cause serious health consequences in IDLH environments for the employee, coworkers, or the public. Job demands such as shift work, irregular meals, erratic exercise, and difficulty maintaining a regular medication schedule detrimentally affect stable glycemic control. Medical risk factors for severe hypoglycemia in ITDM include hypoglycemia unawareness, intensive or poor glycemic control, autonomic neuropathy, and beta blockade.[22]

Hypoglycemia is largely, but not only, a concern for people using insulin. Advances in oral hypoglycemic agents and new evidence that type II diabetics benefit from tighter glycemic control inevitably will lead to more significant hypoglycemic problems for people using only oral agents. Sulfonylurea-induced severe hypoglycemic events occur infrequently—about 2 episodes per 100 patient-years. Glyburide and chlorpropamide produce the highest hypoglycemic rates of oral agents.

Metformin and acarbose should not cause similar problems. The risk for hypoglycemia increases with advanced age, poor nutrition, drug interactions with sulfonylureas, and hepatic or renal disease leading to reduced metabolism or excretion of the drugs.[22]

Questions remain as to whether ITDM actually results in adverse occupational events. Assuming that hypoglycemia results in accidents, the DOT has used old data and models to predict that ITDM heightens a commercial driver's risk of motor vehicle accidents.[24] Empirical data collected by the DOT during its ITDM waiver program and epidemiologic studies of noncommercial drivers provide mixed conclusions about the impact of insulin treatment on accident rates.[25] Driving simulators, however, have demonstrated that driving performance deteriorates significantly when blood glucose drops below 65 mg/dl.[26] Unfortunately, hypoglycemia impairs an individual's judgment about his or her capabilities. At least 50 percent of experimental subjects with diabetes decide to drive at least 50 percent of the time when their blood glucose drops to levels that clearly impair their psychomotor capabilities.[27] Data on other types of occupational accidents not related to motor vehicles remain inconsistent.[28]

Obesity

A National Institutes of Health (NIH) consensus panel identified a continuous direct relationship between body mass index (BMI) and morbidity and mortality as BMI rises above 25.[29] BMI [(body weight in kilograms) ÷ (height in meters) ^{2}] is based on the Metropolitan Life Insurance Company Tables of Heights and Weights. About a third of Americans between 20 and 75 years of age meet this definition of obesity.[30] Individuals with morbid obesity (twice ideal weight or greater than 45.5 kg overweight) are those most likely to have imminent problems with respirator use. They suffer from ventilatory disorders, functional limitations of activities of daily living, decreased life expectancy, increased incidence of sudden death, and increased risk of fatal accidents.[31]

Heat-Related Illnesses

The use of respirators with personal protective clothing can add dramatically to the heat stress associated with a job task. Heat stress may lead to incapacitation (muscle cramping and heat exhaustion) or death (heat stroke). In addition to personal protective equipment (PPE), other occupational risk factors include shift work, lack of acclimatization, and dehydration. Cardiovascular fitness, skin disorders that inhibit sweating, several underlying medical problems (e.g., cardiac disease, peripheral vascular disease, certain neurologic disorders, diabetes), advanced age, obesity, sleep deprivation, many classes of medications, and concurrent acute illness predispose to heat stroke.[32] Repeated episodes of heat-related illnesses raise concerns about an underlying thermoregulatory problem that predisposes to heat illnesses. However, a single incident of heat illness in the past probably does not increase the risk for a recurrent episode.[33,34]

In the absence of other risk factors, cardiovascular fitness, as defined by Vo_{2max}, is a good predictor of heat tolerance. Two individuals with widely different Vo_{2max} values have very different core body temperatures while performing the same workload. Current recommendations for a sustainable workload over an 8-hour shift suggest a limit of 30 to 40 percent of an individual's Vo_{2max}. At 50 percent of Vo_{2max}, rectal temperature rises to 38°C in the absence of heat stress.[35] Clinicians may obtain a worker's Vo_{2max} through simple physiologic testing such as with a bicycle ergometer. Alternatively, at much less cost, physicians and LHCPs can use a model developed by Jackson[36] that considers sex, age, body composition, and self-reported activity to closely predict aerobic functional capacity without exercise testing.

Interview

Diabetes. The PLHCP should interview the employee to gather additional information about the diabetes history (see Figure 4-1). Without a history of complications, referral for further evaluation often will not be necessary. Instead, after obtaining an appropriate release, the physician or LHCP should request the following information from the individual's primary care clinician:

- Type of diabetes: type I (insulin-dependent), type II (non-insulin-dependent), or glucose intolerance (gestational, medication-related)
- Year of initial diagnosis
- Stability of glycemic control over the last year: fasting blood sugars, average blood sugar, hemoglobin A1c, home monitoring results
- Treatment: diet, medications with dosing schedule, any expected changes in treatment
- Insulin dose and schedule, method of administration
- Short-term complications: hypoglycemia (how often, severity,

awareness, and presence of hypoglycemic risk factors), diabetic coma

- Long-term end-organ complications: ocular (cataracts, retinopathy, glaucoma, visual acuity; last eye examination), kidney disease, neuropathy, skin complications, cardiovascular problems
- History of work performance problems related to diabetes
- Assessment of individual's knowledge of the disease and treatment compliance

Obesity. The PLHCP should conduct a review-of-systems interview or probe other questions on the OSHA questionnaire to ascertain that no medical complications accompany the obesity.

Heat-Related Illness. The PLHCP should interview the worker to probe the circumstances that lead to the prior episode(s) of heat-related illness and to identify current risk factors (see Figure 4-1).

- *Occupational risk factors:* environmental conditions, PPE, exertion level, acclimatization, hydration, shift work, duration of work
- *Personal risk factors:* cardiovascular fitness, underlying medical conditions, medications, obesity, and concurrent acute illnesses

Assessment

Diabetes. With a complete absence of research specific to the use of respirators by workers with diabetes mellitus, physicians and LHCPs must depend on their clinical judgment with reference to the general medical and occupational literature as outlined earlier. Stability of glycemic control and the presence of risk factors for hypoglycemia are the prime considerations in respirator approval. The examiner also should identify occupational demands known to disrupt control, such as shift work, irregular diet, illness, and substantial changes in activity levels. The physician or LHCP must consider the hazards associated with the job tasks and work environment in the approval process. In general, persons with diabetes who are managed by diet or oral medications can use respirators safely in most circumstances. Persons with ITDM should show good evidence of glycemic stability, treatment compliance, and understanding of their disease. Great caution should be used in approving persons with ITDM who will work in IDLH environments, especially if that work may involve exposures to factors likely to disrupt glycemic control. Other less hazardous work rarely will require respirator restrictions.

An affirmative answer to any of the following questions may indicate the need for respirator denial or restrictions and should prompt advanced medical evaluation.

- Does the person use insulin?
- Has the person had problems with severe hypoglycemia?
- Does the person have chronic complications of diabetes that may interfere with safe respirator use?
- Does the person have a history of respirator intolerance due to diabetes problems?
- Will the person be working under conditions likely to disrupt glycemic control?
- Does the person show a lack of understanding of his or her disease or compliance with medical recommendations?
- Will the person be working in an IDLH atmosphere?

Obesity. Obesity by itself should never disqualify an individual from work or respirator use. The physician or LHCP, however, should consider a worker's obesity as a reason to probe further about the possibility of heretofore unrecognized medical problems associated with obesity. The Nuclear Employee Data System uses 35 percent over the 1983 Metropolitan Life "ideal weight" as a point to delay respirator qualification pending further evaluation.[6]

An affirmative answer to the following question may indicate the need for respirator denial or restrictions and should prompt the LHCP to refer the examinee to an appropriate clinician:

- Does the worker have diseases associated with obesity covered elsewhere in these guidelines that may interfere with safe respirator use?

Heat-Related Illness. A history of a single episode of heat-related illness need not disqualify a worker from wearing a respirator. Even with the use of full protective clothing and an SCBA, cooling PPE, and reduction of risk factors through physical conditioning, hydration, acclimatization, and rapid personnel rotation can protect most workers. Exceptions may include individuals with an underlying, predisposing medical problem or essential use of medications previously associated with the illness.

An affirmative answer to the following question may indicate the need for respirator denial or restrictions and should prompt advanced medical evaluation.

- Does the worker have risk factors for recurrent heat-related illness?

Plan for Workers with Diabetes

Respirator Approval Decision

1. Diet controlled diabetes without a history of complications
   - Approve for all respirators without additional consultation or evaluation. (A clinical examination is recommended prior to SCBA approval.)
   - Recommend follow-up evaluation at a minimum of an annual basis or earlier if a change in treatment occurs.

2. Diabetes controlled by oral medication:

   (a) Persons *without* complications:

   - Approve for all respirators after receiving information from the treating clinician that documents that the disease is stable and that there are no contraindications to respirator use. (A clinical examination is recommended prior to SCBA approval.)

   (b) Persons *with* complications:

   - Defer respirator approval pending advanced medical evaluation and optimization of therapeutic management.
   - Provide employee and employer with an appropriate timeline for completion of the evaluation and treatment optimization.
   - In either case, recommend follow-up evaluation at a minimum of an annual basis or earlier if a change in treatment occurs.

3. Insulin-treated diabetes mellitus:

   (a) Persons controlled on insulin without complications may be approved for respirator use without restrictions with the following conditions:

   - A clinical examination is recommended prior to SCBA approval.
   - Approve for non-IDLH environment after receiving documentation from treating clinician that diabetes is stable

and that there are no contraindications to respirator use.

- If person will work in an IDLH environment, defer respirator approval pending advanced medical evaluation.
- Recommend follow-up evaluation at a minimum of an annual basis or earlier if a change in treatment occurs.

(b) Persons with diabetes controlled on insulin with complications:

- Defer respirator approval pending advanced medical evaluation and optimization of therapeutic management.
- Provide employee and employer with an appropriate timeline for completion of the evaluation and treatment optimization.
- Recommend follow-up evaluation at a minimum of an annual basis or earlier if a change in treatment occurs.

Additional Considerations for Advanced Respirator Approval
Persons with diabetes and complications:

- Defer respirator approval pending advanced medical evaluation and optimization of therapeutic management.
- Provide employee and employer with an appropriate timeline for completion of the evaluation and review for approval for respirator use.
- Consider obtaining written recommendation from the attending or consulting diabetologist.
- Consider obtaining a record of home monitoring over the past month.
- Consider the hazards of the work environment, the consequences of respirator failure, and the severity and frequency of hypoglycemia, with particular attention to evidence of lack of awareness of hypoglycemic spells and whether they are explained by controllable or predictable events.
- Advanced clinicians should recommend other accommodations and restrictions specific to the work site or recommend denial of respirator approval.
- If the employee's diabetes is likely to interfere with safe use of a respirator while performing assigned tasks, recommend accommodations and restrictions specific to the worksite. If no

accommodations or restriction will make safe use of a respirator possible, recommend denial of respirator approval.

- Individualize recommendations for periodic reevaluations based on job tasks, risk factors, stability and severity of disease, and compliance with medical treatment.

Education

1. Review the impact of work activity and environment on insulin and oral medication needs and risk of hypoglycemia.
2. Emphasize the importance of home monitoring, compliance with medical treatment, and reporting of hypoglycemic events.
3. Describe the need for reevaluation for respirator use in the event of a new treatment regime, difficulties tolerating respirator, or onset of psychomotor, respiratory, cardiac, or neurologic problems.
4. Obtain a release to notify first aid team of worker's condition.
5. Recommend a "buddy system" for ITDM individuals.
6. Encourage pneumovax and annual flu immunization.
7. Report respirator intolerance.

Potential Accommodations for Workers with Diabetes

1. Provide consistent food, shift, and break times.
2. Provide breaks and privacy for administration of medication and self-monitoring.
3. Provide a storage area for medication and supplies.
4. Allow temporary restrictions from all respirator use for planned substantive changes in treatment regime
5. Eliminate job requirement to work in IDLH atmosphere.
6. Eliminate job requirement to work with respirator.

The Job Accommodation Network lists many ideas to help accommodate workers with diabetes.[37]

Plan for Obese Workers

Respirator Approval Decision
Obesity is defined as greater than 35 percent of 1983 Metropolitan Life "ideal weight."

- In the absence of complicating medical conditions, PLHCP may approve for all respirators without additional consultation or evaluation. (A clinical examination is recommended prior to SCBA approval.)
- Recommend reevaluation by brief questionnaire (see Table 3-3) or age-related guideline (see Table 3-4), barring any change in health status or working conditions.
- In the presence of complicating medical conditions, follow relevant guidelines.

Education

1. Advise weight loss.
2. Inform about risk for obesity-associated medical conditions.
3. Advise to report new medical conditions.
4. Report respirator intolerance.

Potential Accommodations for Obese Workers

- Based on accompanying medical conditions.
- Additional protection from heat stress with cooling PPE, rotation, acclimatization, hydration, and physical conditioning.

Plan for Workers with a History of Heat-Related Illness

1. Single episode of heat-related illness in the absence of another identified risk factor for recurrent heat-related illness:
   - Approve for all respirators without additional consultation or evaluation. (A clinical examination is recommended prior to SCBA approval.)
   - Recommend reevaluation by brief questionnaire (see Table 3-3) or age-related guideline (see Table 3-4), barring any change in health status or working conditions.
2. Multiple episodes of heat-related illness or a single episode with risk factors for recurrent heat-related illness but work conditions unlikely to lead to significant heat stress:

- Approve without additional consultation or evaluation unless other conditions are identified that may impair safe use of respirator, in which case relevant guideline should be followed.
- Recommend reevaluation by brief questionnaire (see Table 3-3) or age-related guideline (see Table 3-4), barring any change in health status or working conditions.

Additional Considerations for Advanced Respirator Approval

1. In the presence of risk factors for recurrent heat-related illness and work conditions likely to lead to significant heat stress:
   - Defer respirator approval pending advanced medical evaluation to assess whether risk factors are modifiable.
   - Provide employee and employer with an appropriate timeline for completion of the evaluation and review for approval for respirator use.
   - Consider referring employee for physical conditioning.
   - Ensure routine good occupational medical practices to minimize heat stress.
   - Consider recommending annual reevaluation.
2. If risk factors not sufficiently modifiable to reduce risk:
   - Approve for respirator use in temperate environments but recommend denial of respirator approval for use in hot environments.
   - Recommend annual reevaluation.

Education

1. Advise about medical risk factors.
2. Advise about personal behaviors that can reduce risk, such as hydration, acclimatization, physical conditioning, and adequate sleep and rest.
3. Advise regarding signs of heat illness and first aid.
4. Report respirator intolerance.

Potential Accommodations for Workers with Risk Factors for Heat-Related Illness

- Address any needs presented by accompanying medical conditions.
- Provide additional protection from heat stress with cooling PPE, rotation, acclimatization, hydration, and physical conditioning.
- Eliminate requirement to work in conditions of heat stress.

Respiratory Allergies and Other Immnunologic Disorders

2c. Have you ever had allergic reactions that interfere with breathing?

Forty to fifty million people in the United States have allergies,[38] and 35 million have chronic sinusitis.[39] Over 9 percent of the working-age population of the United States has allergic rhinitis,[39] and approximately another 5 percent of adults are asthmatic.[40] Respiratory allergies can congest the upper respiratory system or lead to pulmonary conditions that interfere with an individual's ability to work safely while wearing a respirator (see questions 3 and 4). Chronic rhinosinusitis can interfere with the ability to detect odors that may provide warnings about respirator failure (see question 2e). Because of its impact on the normal partition between nasal and oral breathing, it also may lead to difficulties with the psychophysiologic aspects of respirator tolerance (see question 8e). The secretions of chronic rhinorrhea may drive a worker to remove a respirator, and cough breaks the facial seal of negative-pressure respirators. Some allergic conditions may require use of medications that interfere with psychomotor functions or whose dosing schedule requires breaks from respirator use for their administration. Physicians and LHCPs must ask the worker who reports respiratory allergies several additional questions to probe whether these very common conditions actually would be an issue for this individual in a particular workplace while wearing a respirator.

In general, although respirators may help protect individuals from sensitization,[41] employers should not rely on them to protect workers already sensitized to an occupational respiratory allergen. Respirators decrease workers' exposure to antigens and can diminish symptoms of occupational asthma.[42] Nonetheless, optimal occupational medicine practice involves complete avoidance of the antigen to which an individual has been sensitized. Seasonal hay fever sufferers, allergic to pollens, probably will not have problems indoors while wearing a respirator because they will not be exposed to the provoking allergen. On the other hand, a person with severe asthma precipitated by grass pollen should not be given a respirator with the expectation that it will provide adequate protection. Respirators are obviously useless against dermal contactants, which occasionally can produce systemic allergic reactions.[43] Examples include latex gloves and the silicone rubber of a respiratory protection device.

Interview

The physician or LHCP should interview the employee to gather additional information about the allergic history (see Figure 4-1). Limited, episodic allergic reactions not likely to occur in the workplace rarely will require referral for further evaluation. Chronic conditions or those involving sensitization to occupational allergens to which an individual may be exposed require referral. Collect the following information by interview or, if necessary, from the individual's treating clinician:

- Allergens to which the individual is sensitized
- Symptoms (eye irritation, wheezing, cough, congestion, i.e., nasal, sinus, or lung, shortness of breath, diminished olfactory acuity) and allergic diagnosis, if available
- Temporal considerations (chronic versus episodic, seasonal, activity-related, or environment-related)
- Diagnostic testing, especially skin test and RAST results
- Treatment and dosing schedule of medications
- Effectiveness of treatment in controlling symptoms
- History of respirator intolerance

Assessment

Most allergic conditions do not interfere with safe respirator use. Critical issues in the clinical assessment are whether allergic problems will manifest in the workplace and result in respirator failure, physiologic intolerance, or reactions to inadvertent exposure to offending antigens. An affirmative answer to any of the following questions may indicate the need for respirator denial or restrictions and should prompt more advanced medical evaluation.

- Are the allergic symptoms likely to provoke a break in the facial seal of a negative-pressure respirator?
- Are the allergic symptoms likely to provoke noncompliance with respirator use? Is the worker likely to remove the respirator to

wipe clean nasal secretions or administer medications?

- Is the nasal congestion sufficient to interfere with odor detection in circumstances where the worker must rely on odor as a warning of exposure (question 2e)?
- Are the allergic symptoms sufficient to place substantial physiologic limitations on the person's ability to tolerate a respirator (questions 3 and 4)?
- Does the person use medications likely to affect psychomotor functions (question 7a)?
- Does the person have a history of respirator intolerance due to allergy problems?
- Will the person be working with substances known to provoke respiratory or other systemic reactions in this individual?
- Will the person be working in an IDLH atmosphere?

Plan

Respirator Approval Decision

1. Individuals who report that their allergic respiratory reactions are provoked only to a food substance, insect venom, or respiratory allergen not normally uniquely encountered in the person's work:
   - Approve for all respirators without additional consultation or evaluation. (A clinical examination is recommended prior to SCBA approval.)
   - Recommend reevaluation by brief questionnaire (see Table 3-3) or age-related guideline (see Table 3-4), barring any change in health status or working conditions.
2. Individuals who have an allergic condition that leads to affirmative answers to the assessment considerations listed above:
   - Defer respirator approval pending advanced medical evaluation and optimization of therapeutic management.
   - Provide employee and employer with an appropriate timeline for completion of the evaluation and treatment optimization.
   - Base recommendation for periodic reevaluation on results of abdominal medical evaluation.

3. Individuals with impaired sense of smell: See protocol for question 2e.

Additional Considerations for Advanced Respirator Approval

1. Individuals with poorly controlled symptoms likely to result in respirator intolerance:
   - Defer respirator approval pending advanced medical evaluation and optimization of therapeutic management.
   - Provide employee and employer with an appropriate timeline for completion of the evaluation and review for approval for respirator use.
   - Consider the hazards of the work environment, the consequences of respirator failure, the severity and frequency of allergic respiratory symptoms, and whether they are explained by controllable or predictable events.
   - Consider recommending use of a specific respirator such as a powered air-purifying respirator (PAPR) to resolve concerns about noncompliance, breaks in respirator seal, or physiologic effort.
   - If the employee's allergic condition or medications is likely to interfere with safe use of a respirator while performing assigned tasks, recommend accommodations and restrictions specific to the worksite. If no accommodations or restriction will make safe use of a respirator possible, recommend denial of respirator approval.
   - Individualize recommendations for periodic reevaluations based on job tasks, risk factors, stability and severity of disease, and compliance with medical treatment.
2. Individuals using medications likely to affect psychomotor function:
   - Coordinate changes in medication (type, dosage, and schedule) with treating physician.
   - Consider the hazards of the work environment and the consequences of errors in judgment, alertness, coordination, or other psychomotor functions.
   - Assess psychomotor function by asking about any self-perceived effects and consider review of job performance in a

non-IDLH atmosphere.

Education

1. Report symptoms or use of medications that may temporarily interfere with safe work with a respirator.
2. Report respirator intolerance.
3. Report change in health status.
4. Suggest follow-up with an appropriate clinician if symptoms are not adequately controlled.

Potential Accommodations for Workers with Allergic Reactions that Interfere with Breathing

1. Offer a storage area for medications.
2. Allow temporary restrictions from respirator use during allergic exacerbations or intermittent use of medications that affect psychomotor function.
3. Provide PAPRs to address problems with respirator tolerance, seal, and physiologic effort.
4. Eliminate job requirement to work in an IDLH atmosphere.
5. Eliminate job requirement to work with a respirator.

Psychological Conditions

2d. Have you ever had claustrophobia (fear of closed-in places)?

Psychophysiologic factors probably represent the most common reasons for respirator intolerance. Because of the confining nature of respirators, claustrophobia typifies the most obvious psychological condition that may predispose to respirator failure. Anxiety or panic, for any reason, may drive a person to inappropriately remove a respirator. If this occurs in an IDLH atmosphere, serious harm may result. Similar reactions may occur in individuals with other psychological diagnoses, such as panic disorder or generalized anxiety disorder. In addition, psychological conditions associated with respirator intolerance may lead to significant elevations in heart rate, blood pressure, and respiratory rate that can jeopardize the health of employees who are at high risk for cardiopulmonary disease.[44]

The medical evaluation for respirator use generates a work-related medical record that belongs to employers. Because of the sensitive nature of psychological diagnoses and their associated clinical information, the clinician in this setting should rarely probe beyond simple questions about medications and behavior directly relevant to respirator use in hazardous environments.

Some psychological conditions may require use of medications that interfere with psychomotor functions or whose dosing schedule requires breaks from respirator use for their administration. Physicians and LHCPs must ask the worker who reports claustrophobia or other psychological conditions several additional questions to probe whether this individual will have problems in a particular workplace while wearing a respirator.

Interview

The physician or LHCP should interview the employee to gather additional information about the history of claustrophobia (see Figure 4-1). The physician or LHCP also should ask open-ended questions that may yield other psychiatric information that may predict respirator intolerance. At the same time, the examiner should not cast a wide net to obtain information about psychiatric conditions unlikely to be relevant to the use of a respirator. Most people reporting claustrophobia, anxiety, or other psychological conditions will not require restrictions or respirator denial. Individuals who will require restrictions probably will state their fears of respirator use quite clearly, they will report past problems with respiratory tolerance, or their problems will manifest themselves during fit testing. As with other medical conditions, past successful respirator use in similar environments best predicts future successful use. Collect the following information by interview or, if necessary, from the individual's treating clinician:

- History of respirator tolerance or intolerance
- History of severity of anxiety and panic attacks, with specific inquiry about behavior during an episode
- Symptoms and behaviors likely to affect safe respirator use, such as poor judgment, hallucinations, active suicidal ideation, psychomotor impairment
- Specific environments or other circumstances that precipitate anxiety
- Examinee's own prediction of ability to tolerate respirator in the

proposed work environment

- Treatment and dosing schedule of medications
- Effectiveness of treatment in controlling symptoms

Assessment

Various studies have documented that 20 to 25 percent of the population have significant emotional symptoms.[45] Few of these people, however, will require respirator denial or restriction. More often, psychophysiologic factors resulting in respirator noncompliance are not associated with specific psychological diagnoses but are unique to respirator use. For this reason, the most critical piece of information, when available, is past history of respirator tolerance. An affirmative answer to any of the following questions may indicate the need for respirator denial or restrictions and should prompt advanced medical evaluation.

- Does the worker use medications (question 7a) or have an uncontrolled condition that affects psychomotor functioning or is associated with behavior incompatible with safe respirator use or the job tasks themselves?
- Has the worker's symptoms led to uncontrollable urges to leave environments similar to the work setting?
- Does the worker have a history of respirator intolerance or intolerance of environments whose features are similar to those at the proposed work site?

Plan

Respirator Approval Decision

1. Individuals with active claustrophobia or other relevant psychological distress but able to wear similar respirators in the past while performing similar job tasks without problems and denying concerns about respirator tolerance:
    - Approve for all respirators without additional consultation or evaluation. (A clinical examination is recommended prior to SCBA approval)
    - If respirator type is significantly different from past use, e.g., dust mask versus full-face respirator, assessment for claustrophobia during fit test and/or work trial may be

appropriate.

- Arrange work trial if tolerance is questionable.
- Recommend reevaluation by brief questionnaire (see Table 3-3) or age-related guideline (see Table 3-4), barring any change in health status or working conditions.

2. Individuals who report a history or fear of respirator intolerance:
   - Defer respirator approval pending advanced medical evaluation and optimization of therapeutic management.
   - Provide employee and employer with an appropriate timeline for completion of the evaluation and review for approval for respirator use.
   - If the individual refuses referral or work trial, deny approval or restrict respirator use without additional consultation.
   - Base recommendations for periodic reevaluation on results of advanced medical evaluation.
3. Individuals who have other psychiatric conditions that lead to affirmative answers to the assessment considerations listed above:
   - Defer respirator approval pending advanced medical evaluation and optimization of therapeutic management.
   - Provide employee and employer with an appropriate timeline for completion of the evaluation and review for approval for respirator use.
   - Base recommendations for periodic reevaluation on advanced medical evaluation.

Additional Considerations for Advanced Respirator Approval

1. Individuals with poorly controlled symptoms likely to result in respirator intolerance:
   - Defer respirator approval pending advanced medical evaluation and optimization of therapeutic management.
   - Provide employee and employer with an appropriate timeline for completion of the evaluation and review for approval for respirator use.
   - Consider the hazards of the work environment, the

consequences of respirator failure, and the severity and frequency of psychiatric symptoms and whether they are explained by controllable or predictable events.

- Consider use of a loose-fitting respirator such as PAPR that may decrease claustrophobia.
- Offer a work trial.
- If the employee's psychological condition or medical condition or medication use is likely to interfere with safe use of a respirator while performing assigned tasks, recommend accommodations and restrictions specific to the worksite. If no accommodations or restriction will make safe use of a respirator possible, recommend denial of respirator approval.
- Individualize recommendations for periodic reevaluations based on job tasks, risk factors, stability and severity of disease, and compliance with medical treatment.

2. Individuals using medications likely to affect psychomotor function:
    - Defer respirator approval and coordinate possible changes in medication (type, dosage, and schedule) with treating physician.
    - Provide employee and employer with an appropriate timeline for completion of the evaluation and review for approval for respirator use.
    - Consider the hazards of the work environment and the consequences of errors in judgment, alertness, and coordination as well as other psychomotor functions.
    - Assess psychomotor function by asking the worker about any self-perceived effects and consider review of job performance in a non-IDLH atmosphere.

Education

1. Report symptoms or use of medications that may temporarily interfere with safe work with a respirator.
2. Report respirator intolerance.
3. Report change in health status.
4. Suggest follow-up with a more appropriate clinician if symptoms

are not adequately controlled.

Potential Accommodations for Workers with Claustrophobia or Other Psychological Conditions that Interfere with Respirator Tolerance

1. Offer PAPR or other loose-fitting respirator.
2. Allow temporary restrictions from respirator use during exacerbations of psychological condition or use of medications that deleteriously but temporarily affect psychomotor function.
3. Eliminate job requirement to work in an IDLH atmosphere or specific environments that provoke anxiety.
4. Eliminate job requirement to work with a respirator.

The Job Accommodation Network lists many ideas to help accommodate workers with psychiatric disorders.[46]

Olfactory Disorders

2e. Have you ever had trouble smelling odors?

Olfactory dysfunction afflicts at least 2 million people in the United States.[47] About 2 percent of the working population suffers from a smell disorder.[48] Problems range from complete absence of smell (anosmia), to decreased sense of smell, to distorted smell (dysosmias). Olfaction provides workers with a valuable warning system about exposure to potentially hazardous agents. One of OSHA's qualitative fit-testing protocols relies on the respirator user's sense of smell. In some cases, odor may trigger a worker to don a respirator to investigate an unexpected chemical release or to escape an environment. The Nuclear Regulatory Commission, for example, considers the ability to smell combustion products and marker gases an important physical qualification for licensed operators of nuclear power plants.[7] An operator who identifies such odors in the control room of a nuclear power plant dons a respirator and conducts certain procedures to ensure the safety of the reactor. Similarly, NFPA 1582 recommends that anosmia disqualify a candidate for firefighting.[5]

Odor detection also may alert a worker to respirator failure. End-of-service-life indicators and scheduled cartridge replacement mandated by the new OSHA respiratory protection standard reduce but do not eliminate the role of odor detection to identify spent air-purification media. Since the sense of smell can alert a worker to poor respirator fit, failures in respirator technology, and the need to don a respirator, olfactory disturbances may affect the safety of a respirator user.

Individuals with olfactory impairment may exhibit hyposmia, a decreased ability to smell some or all odors; cacosmia, a distortion of the sense of smell; or anosmia, the inability to smell any odors. Transient reductions in the sense of smell are extremely common sequelae of allergic or infectious conditions. Disturbed olfaction may persist because of congenital abnormalities, aging, chronic rhinosinus disease, the sequelae of an acute upper respiratory infection, or head trauma. No cause can be found in between 19 and 26 percent of patients with olfactory dysfunction. Another 20 percent have underlying medical conditions secondarily affecting the sense of smell, including endocrine, neurologic, psychiatric, and nutritional disorders.[47] In some cases, medical treatment can restore normal olfaction.

Workers sometimes can rely on other senses to provide sufficient environmental information to provoke appropriate behavioral responses. The common chemical sense, mediated by the trigeminal nerve in the nose, can provide warnings about irritant exposures. Other mucous membranes, especially the eyes, are sensitive to irritant exposure. Vision, hearing, and even taste similarly sometimes may provide the information necessary to alert a worker about a hazardous exposure. A person may see or hear a fire or an unexpected chemical release. Auditory or visual alarms also may signal hazardous emissions requiring respirator use.

Interview

The physician or LHCP should interview the employee to gather additional information about the olfactory disturbance (see Figure 4-1). The physician or LHCP should clarify whether the condition persists or was only transiently associated with nasal obstruction. The clinician also should ascertain whether the condition represents true anosmia or, less significantly, hyposmia and whether the worker has ever been evaluated and treated for this problem. Detailed knowledge about the potential occupational exposures is essential to assess the significance of olfactory disorders. Collect the following information by interview or, if necessary, from the individual's treating clinician:

- Date of onset
- History of evaluation
- Etiology, if known

- Treatment
- Severity of the disorder
- Whether the worker reports an inability to smell the types of odors produced by expected hazards in the proposed workplace

The physician or LHCP should perform and interpret olfactory acuity screening (see Chapter 3) to collect empirical information if the worker reports that the condition persists. If the individual fails the olfactory acuity screening test, the physician or LHCP should conduct a more comprehensive diagnostic test (see Chapter 3). Interpretation of abnormal results requires referral to an appropriate advanced clinician.

Assessment

Detailed knowledge about potential occupational exposures is essential to assess the significance of olfactory disorders. Of central importance is whether the individual can detect the presence of the specific hazardous exposures of concern. If the worker is hyposmic, can he or she smell the substance of concern? The person may not smell the difference between vanilla and apple spice but can readily identify low concentrations of xylene. If the worker is anosmic, does the potential exposure have adequate irritant or gustatory warning properties? The irritant properties of ammonia or chlorine supersede their odorant properties. Gustation can reveal the presence of some airborne hazardous substances. In other circumstances, olfaction may be irrelevant to the hazard, such as the health care worker wearing a respirator for protection from tuberculosis. An affirmative answer to any of the following questions may indicate the need for respirator denial or restrictions and should prompt advanced medical evaluation.

- Does the individual have persistent anosmia?
- Does the individual fail the olfactory acuity screening and more advanced test?
- Does the individual report difficulty smelling the specific hazardous substances in the workplace?

Plan

Respirator Approval Decision

1. Individuals who are able to correctly recognize the odors in the olfactory acuity screening test:

- Approve for all respirators without additional consultation or evaluation. (A clinical examination is recommended prior to SCBA approval.)
- Recommend reevaluation by brief questionnaire (see Table 3-3) or age-related guideline (see Table 3-4), barring any change in health status or working conditions.

2. Individuals who fail the olfactory acuity screening test:
   - Perform more comprehensive testing such as the Sensonics Smell Identification Test.[1]
   - If the person fails the more advanced olfactory test, defer respirator approval pending advanced medical evaluation and optimization of therapeutic management.
   - Provide employee and employer with an appropriate timeline for completion of the evaluation and review for approval for respirator use.
3. Individuals with well-documented persistent anosmia or trouble smelling relevant hazardous substances:
   - Defer respirator approval pending advanced medical evaluation and optimization of therapeutic management.
   - Provide employee and employer with an appropriate timeline for completion of the evaluation and review for approval for respirator use.

Additional Considerations for Advanced Respirator Approval
Individuals with anosmia or hyposmia that may result in the inability to detect relevant substance by olfactory cues:

- Consider the hazards of the work environment and the consequences of an inability to detect respirator failure or unexpected hazardous exposure.
- Consider whether other senses could provide adequate warning information.
- Consider designing a work trial that specifically tests an individual's ability to detect the presence of the hazard of concern. Such a trial must control exposures to minimize the possibility of overexposure (see Chapter 3).

[1]Available from Sensonics, Inc., Haddon Heights, New Jersey.

- Consider advising the employer representative to (1) choose a qualitative fit-testing protocol that does not rely on the sense of smell, (2) perform quantitative fit testing, and (3) offer a loose-fitting respirator that does not require fit testing.
- Recommend accommodations and restrictions specific to the work site or, if none are feasible, recommend denial of respirator approval.
- Individualize recommendations for periodic reevaluation, based on job tasks, risk factors, stability and severity of disease, and compliance with medical treatment.

Education

1. Report transient changes in olfactory acuity critical to safe job performance.
2. Report changes in the work environment that may require change in accommodations.

Potential Accommodations for Workers with Olfactory Disorders

1. Choose a qualitative fit-testing protocol that does not rely on the sense of smell, perform quantitative fit testing, or provide a loose-fitting respirator that does not require fit testing.
2. Provide a colleague with a normal sense of smell who can alert the anosmic individual to dangerous environmental conditions.
3. Allow temporary restrictions from respirator use during transient decreases in essential olfactory acuity.
4. Identify a method for detecting the presence of a hazardous substance by a means other than olfaction.
5. Eliminate job requirement to work in an IDLH atmosphere or specific environments that contain hazardous substances that the individual cannot smell or otherwise detect.
6. Eliminate job requirement to work with a respirator.

Pulmonary Disorders and Symptoms

3. Have you ever had any of the following pulmonary or lung problems (a–l)?

4. Do you currently have any of the following symptoms of pulmonary or lung illness (a–n)?

Questions 3 and 4 of the OSHA Respirator Medical Evaluation Questionnaire address pulmonary diagnoses, severity of pulmonary symptoms, and nonspecific symptoms often associated with pulmonary disease. The many conditions and symptoms addressed by these questions resolve into the following general concerns: pulmonary function (questions 3a–d, g). respiratory symptoms without diagnosis (questions 4a–n), infection (questions 3e–f), and some miscellaneous conditions (questions 3h–l). The organization of this section of the guidelines follows these categories rather than slavishly paralleling the OSHA questions.

Disorders of Pulmonary Function

Many of the conditions identified in the OSHA questionnaire are relevant to respirator use because of their effect on pulmonary function. Respirators do provide a physiologic load on the pulmonary system. Laboratory-based experiments led to the historical assumption that ventilatory limitations identified by spirometry would predict respirator tolerance.[3] At submaximal effort levels, the physiologic effects resulting from respirator use, however, have been shown to be small and generally well tolerated by both healthy individuals and those with impaired lung function.[2]

Commonly, pulmonary function values were used as a litmus test for respirator certification. Current knowledge has established that pulmonary function testing by itself is neither a sensitive nor a specific indicator for respirator tolerance or safety. Physicians and LHCPs therefore should *not* rely solely on pulmonary function or diagnoses as valid predictors of respirator tolerance.[49] The ATS points out that measurement of ventilatory function nevertheless often may be useful in assessing a worker's general work capacity.[2] Answers to questions about how shortness of breath or wheezing affects activities of daily living or occupational function provide rough indicators of a person's ability to perform specific work tasks.

Spirometric screening reliably identifies the fixed obstructive or restrictive dysfunction associated with pneumoconioses and chronic obstructive pulmonary disease (COPD). Periodic reevaluation identifies progression of ventilatory problems. Simple spirometry may understate the severity of respiratory disease. In selected cases, more advanced pulmonary testing, including complete pulmonary function testing with diffusion capacity, arterial blood gasses, and cardiopulmonary stress testing, may be necessary to better characterize the extent of the disease. COPD also may have a reversible component of obstruction exacerbated by infection.

Documentation of reversible obstruction is an essential diagnostic criterion of asthma. Routine spirometry often will not identify the abnormalities associated with intermittent asthma. Questionnaire, interview, and medical records provide tools that are more sensitive for detecting asthma. Repeated spirometry, before and after bronchodilation, along with methacholine challenge and ongoing peak flow monitoring, may better characterize the severity of the disease. Variability in airflow, the hallmark of asthma, presents the biggest challenge in assessing a worker's ability to work safely with a respirator.

The additional psychophysiologic load of a respirator may lead to respirator noncompliance in a worker with compromised pulmonary function. Further decreases in aerobic capacity may result in critical job failure in high-demand circumstances. Increased ventilatory effort, adverse thermal environments, and inadvertent exposure to respiratory irritants or allergens may provoke an asthma attack. Respirators can decrease antigen exposure and consequently, at least in the short term, reduce associated asthmatic symptoms.[42] Unfortunately, respirators cannot fully protect asthmatics from exposure to respiratory antigens.[50] Not only can such exposure precipitate acute asthmatic attacks, but it also can lead to increasing sensitization. Therefore, respirator use for the purpose of continued work with substances to which an individual has significant clinical reactions is strongly discouraged except when accompanied by careful risk communication and ongoing pulmonary monitoring. This use of respirators should be reserved for situations where an individual's livelihood, such as family farming, cannot otherwise be preserved. Workers with occupational lung diseases, such as asbestosis or hypersensitivity pneumonitis, are also at risk of aggravating their disease with further exposure to the offending agent and should in general be removed from potential exposure. Evaluation for respirator use for individuals with pulmonary dysfunction must focus on the severity of their disease as measured by symptoms, activity tolerance, dependence on medication, triggers of exacerbations, and the nature of the work in which the respirator will be used.

Respiratory Symptoms without Diagnosis

Workers with undiagnosed chronic respiratory symptoms require appropriate evaluation and treatment before completing the respiratory qualification process. Regardless of the diagnosis, the symptoms themselves may interfere with respirator tolerance. Pleuritic pain restricts ventilatory effort, reducing work capacity and ventilation and increasing the possibility of respirator noncompliance. Most individuals with pleuritic pain have an acute disease or injury, such as chest wall trauma or pleurisy, that will resolve. Chronic pain with respiration may be symptomatic of a serious underlying condition that also will impair an individual's respirator and work tolerance. Chronic cough will break the facial seal of tight-fitting respirators. The problem becomes more significant when the cough produces sputum.

Infection

Active respiratory infections provoke symptoms and generalized debility that impair safe use of a respirator. Depending on their severity, upper respiratory infections may interfere with safe respirator use. Uniformly, acute bronchitis, pneumonia, and tuberculosis require treatment and resolution.

Miscellaneous Conditions

Several other conditions, some identified on the OSHA questionnaire, potentially may affect respirator use. Generally, the limiting factor is not the diagnosis but rather accompanying problems with pulmonary function, respiratory symptoms, and general debility. Lung cancer, for example, need not itself be disqualifying, but associated surgery, chemotherapy, and radiotherapy may well be. Fractured ribs heal, but if they represent a pathologic fracture, the underlying disease may disqualify the person for respirator use.

Chest surgery may result in temporary impairment due to pleuritic pain or permanent reductions in lung function due to lobectomy. Even after pneumonectomy, rehabilitation may lead to a satisfactory capacity for respirator tolerance. Pulmonary function, underlying diagnosis, and symptoms provide clues to whether the individual who has recovered from surgery will be able to use a respirator safely.

A history of spontaneous pneumothorax traditionally has been listed as a contraindication for respirator use.[51] Although without empirical validation, the reasoning behind this judgment stems from the associated hypoxemia, acute pleuritic pain, and high recurrence rate. Pneumothorax associated with an underlying pulmonary disease generally has a more debilitating effect than primary spontaneous pneumothorax. The latter occurs most commonly in healthy men between the ages of 20 and 40. The relative risk of spontaneous pneumothorax rises astronomically with the numbers of cigarettes smoked daily. A male smoking more than 22 cigarettes a day increases his risk 102 times.[52] Longitudinal studies over 6 years have documented ipsilateral recurrence as high as 52 percent at a mean interval of 2.3 years.[53] Each subsequent pneumothorax raises the risk of recurrence, which is not affected by treatment, with the exception of chemical or surgical pleurodesis. Pleurodesis decreases the risk of recurrence by about half. Nonetheless, the immediate factors that lead to bleb rupture remain unknown. Potential explanations include changes in airway caliber that lead to check-valving of airflow, abrupt variations in atmospheric pressure, and other forces applied to the visceral pleura. Spontaneous primary pneumothorax usually occurs at rest, with less than 10 percent occurring during exercise.[52,54]

Interview

The physician or LHCP should interview the employee to gather additional information about the pulmonary condition or symptoms (see Figure 4-1). The examiner should perform or obtain recent spirometry. Other laboratory investigations, including chest x-rays, require the involvement of an appropriate clinician. After obtaining an appropriate release, the physician or LHCP should procure relevant medical records from the individual's treating practitioner. If the worker has symptoms but no diagnosis or therapeutic plan, the physician or LHCP must refer the individual to an appropriate clinician. Many individuals with pulmonary symptoms are undertreated. Referral for more intensive treatment may resolve any concerns regarding use of a respirator. Despite the wide variety of potential conditions, a single generic set of questions serves to outline the critical data needed to proceed to the assessment phase.

- Date of symptom onset and diagnosis
- Episodic or chronic symptoms
- Frequency of episodes over the past 2 years if episodic, with date of last episode
- Current symptoms
- Factors that trigger exacerbations
- Results and date of relevant diagnostic tests

- Medications and other treatments, including hospitalizations
- Functional impairment, including existing occupational restrictions, history of lost work due to pulmonary illness, impact on exercise or other strenuous activities, impact on activities of daily living
- History of respirator use and tolerance

Assessment

The wide variety of pulmonary conditions requires the physician or LHCP to look beyond the diagnosis to carefully ascertain functional capacity and factors that trigger respiratory deterioration. Workers with stable pulmonary conditions who can perform the essential functions of a job without a respirator most likely will tolerate a respirator while performing the same job. On the other hand, accommodations may be required to circumvent problems, such as cough, that are incompatible with tight-fitting respirators. Workers with adequate pulmonary function but with occupational diseases related to a specific agent may be approved for respirator use but in general should be restricted from further potential exposure to this agent. We urge that workers actively participate in this decision making. Transient conditions that produce symptoms and physiologic impairment, such as respiratory infections, without chronic sequelae require only temporary restrictions from respirator use. The heterogeneity of asthma severity and triggers demands careful individualization of physician or LHCP recommendations. In general, workers should not rely on respirators to protect them from antigens to which they are sensitive.

An affirmative answer to any of the following questions may indicate the need for respirator denial or restrictions and should prompt referral for advanced medical evaluation.

- Are the pulmonary symptoms likely to provoke a break in the facial seal of a tight-fitting respirator?
- Are the pulmonary symptoms likely to provoke noncompliance with respirator use? Is the worker likely to remove the respirator to dispose of sputum, to administer medications, or to decrease respiratory load?
- Are the pulmonary symptoms sufficient to place substantial physiologic limitations on the person's ability to perform activities similar to those which will be required while wearing a respirator?

- Does the individual have an FEV_1 or FVC of less than 70 percent of predicted?
- Does the person use medications likely to affect psychomotor functions (question 7a)?
- Does the person have a history of respirator intolerance due to pulmonary problems?
- Does the person have an episodic condition that could temporarily interfere with safe job performance while wearing a respirator?
- Will the person be working with substances responsible for his or her disease or known to provoke respiratory or other systemic reactions in this individual?
- Will the person with respiratory symptoms be working in an IDLH atmosphere?

Plan

Respirator Approval Decision

Disorders of Pulmonary Function (questions 3a–d,g). Emphysema, chronic bronchitis, pneumoconioses, asthma, restrictive diseases:

1. Individuals with stable restrictive or fixed obstructive disease unrelated to the substances in the work environment and with demonstrated functional capacity to perform job tasks similar to those which will be performed while wearing a respirator:
   - Approve for all respirators without additional consultation or evaluation. (A clinical examination is recommended prior to SCBA approval.)
   - Arrange a work trial if tolerance is questionable.
   - Recommend reevaluation by brief questionnaire (see Table 3-3) or age-related guideline (see Table 3-4), barring any change in health status or working conditions.
2. Individuals with a disorder of pulmonary function that leads to an affirmative response to the assessment considerations listed above:
   - Defer respirator approval pending advanced medical evaluation and optimization of therapeutic management.

- Provide employee and employer with an appropriate timeline for completion of the evaluation and review for approval for respirator use.

Additional Considerations for Advanced Respirator Approval

Chronic Obstructive or Restrictive Pulmonary Diseases

- If the worker is symptomatic or functionally impaired and questions remain about whether that individual is optimally medically managed, defer respirator approval pending advanced medical evaluation and optimization of treatment.
- Provide employee and employer with an appropriate timeline for completion of the evaluation and review for approval for respirator use.
- Consider the hazards of the work environment, the consequences of respirator or job performance failure, the history of respirator tolerance, and the functional impact of disease.
- If required functional capacity is demonstrated without respirator, approve for respirator use without restrictions.
- If capacity to perform similar job tasks has not been demonstrated, arrange for a work trial and base decision on tolerance of work trial.
- If the employee's pulmonary impairment or symptoms are demonstrated to interfere with safe use of a respirator while performing assigned tasks, recommend accommodations and restrictions specific to the work site. If no accommodations or restriction will make safe use of a respirator possible, recommend denial of respirator approval.
- If the employee has a pneumoconiosis, recommend avoidance of the offending agent, but with careful risk communication to include the employee in decision making.
- If the employee has a history of hypersensitivity pneumonitis, with few exceptions, recommend avoidance of offending antigen and denial of respirator use for this purpose.
- Consider recommending reevaluation based on severity and progression of disease as well as confidence in initial recommendations.

Asthma

- If the worker is symptomatic or functionally impaired and questions remain about whether the individual is optimally medically managed, defer respirator approval pending advanced medical evaluation and optimization of treatment.
- Provide employee and employer with an appropriate timeline for completion of the evaluation and review for approval for respirator use.
- Consider the hazards of the work environment, the consequences of respirator or job performance failure, the history of respirator tolerance, the functional impact of disease, and whether asthma triggers are present at work site.
- Consider recommending against asthmatic workers relying on a respirator for protection from antigens to which they are sensitive.
- Consider arranging a work trial if tolerance is questionable and concern relates to response to exertion, environmental temperature, chronic symptoms, or psychophysiologic tolerance. A work trial will not resolve concerns regarding episodic triggers or involvement in response to workplace emergencies.
- If employee's pulmonary impairment or symptoms are demonstrated to interfere with safe use of a respirator while performing assigned tasks or transient exacerbations of the condition will unpredictably affect respirator tolerance, recommend accommodations and restrictions specific to the work site. If no accommodations or restriction will make safe use of a respirator possible, recommend denial of respirator approval.
- Recommend reevaluation based on severity and progression of disease as well as confidence in initial recommendations.

Education

1. Report for reevaluation for respirator use for problems tolerating respirator or performing job.
2. Decline use of respirator for temporary exacerbations such as an asthma attack or respiratory infection.
3. Consider referral to a pulmonary rehabilitation or asthma education program to increase an individual's ability to manage and cope with the disease.

4. Refer for smoking cessation when relevant.
5. Counsel about importance of avoiding asthma triggers at home as well as at work.
6. Encourage pneumovax and annual flu immunization.

Potential Accommodations

1. Provide a PAPR to decrease the work of breathing.
2. Provide a loose-fitting respirator to accommodate symptoms, such as a chronic cough, that would interfere with the facial seal of a tight-fitting respirator.
3. Allow temporary restrictions from respirator use during transient decreases in pulmonary function.
4. Eliminate job requirement to work on a hazardous response team, in an IDLH atmosphere, or in specific environments that contain triggers that would aggravate the respiratory disease if the respirator were removed.
5. Eliminate job requirement to work with a respirator.

Respiratory Symptoms (questions 4a–n). Shortness of breath, productive cough, early morning cough, cough when lying down, coughing up blood, wheezing, chest pain when breathing deeply, any other possible lung symptoms:

1. Individuals with a past history of transient respiratory symptoms not associated with an underlying chronic disease such as asthma or pneumothorax:
   - Approve for all respirators without additional consultation or evaluation. (A clinical examination is recommended prior to SCBA approval.)
   - Recommend reevaluation by brief questionnaire (see Table 3-3) or age-related guideline (see Table 3-4), barring any change in health status or working conditions.

2. Individuals with active minor respiratory symptoms consistent with an upper respiratory infection that probably will interfere with respirator use:
    - Defer respirator approval pending reevaluation by interview in 2 weeks.
    - If symptoms do not resolve, refer for advanced medical evaluation.
    - Provide employee and employer with an appropriate timeline for completion of the evaluation and review for approval for respirator use.
    - If symptoms have resolved, approve for respirator use without restrictions and without additional consultation. (A clinical examination is recommended prior to SCBA approval.)

Additional Considerations for Advanced Respirator Approval

Persistent Respiratory Symptoms without a Diagnosis

- Defer respirator approval pending advanced medical evaluation and optimization of therapeutic management.
- Provide employee and employer with an appropriate timeline for completion of the evaluation and review for approval for respirator use.
- Follow the relevant guidelines after diagnosis known.

Education

1. Emphasize importance of timely diagnostic evaluation for persistent respiratory symptoms of unknown cause.
2. Refer for smoking cessation when relevant.
3. Report respirator intolerance.

Potential Accommodations

1. Allow temporary restrictions from respirator use pending completion of diagnostic evaluation and initiation of treatment.
2. Eliminate respirator use requirement.

Respiratory Infection (questions 3e–f). Pneumonia, tuberculosis:

1. Individuals with a remote (>1 month) history of an acute infection such as pneumonia, bronchitis, or sinusitis without sequelae:
   - Approve for all respirators without additional consultation or evaluation. (A clinical examination is recommended prior to SCBA approval.)
   - Recommend reevaluation by brief questionnaire (see Table 3-3) or age-related guideline (see Table 3-4), barring any change in health status or working conditions.
2. Individuals without respiratory symptoms with a history of a positive PPD test who have been evaluated for prophylactic treatment:
   - Approve for all respirators without additional consultation or evaluation. (A clinical examination is recommended prior to SCBA approval.)
   - Recommend reevaluation by brief questionnaire (see Table 3-3) or age-related guideline (see Table 3-4), barring any change in health status or working conditions.
3. Asymptomatic individuals with a history of a positive PPD test who have not been evaluated for prophylactic treatment:
   - Refer for advanced medical evaluation.
   - Without awaiting recommended evaluation, approve for all respirators. (A clinical examination is recommended prior to SCBA approval.)
   - Recommend reevaluation by brief questionnaire (see Table 3-3) or age-related guideline (see Table 3-4), barring any change in health status or working conditions.
4. Individuals with a history of active tuberculosis:
   - Defer respirator approval pending medical evaluation and optimization of therapeutic management.
   - Provide employee and employer with an appropriate timeline for completion of the evaluation and review for approval for respirator use.
   - Obtain documentation from the treating clinician.

- Defer recommendation for periodic reevaluation on results of advanced medical evaluation.

Additional Considerations for Advanced Respirator Approval

Individuals with Chronic Respiratory Symptoms and a History of a Positive PPD Test

- Defer respirator approval pending advanced medical evaluation to rule out active tuberculosis and to assess and treat nontubercular respiratory symptoms.
- Provide employee and employer with an appropriate timeline for completion of the evaluation and review for approval for respirator use.
- For subsequent to return to work after resolution of contagious state, approve for all respirator use unless sequelae interfere with work capacity or respirator tolerance.
- Follow relevant guidelines protocol if another diagnosis for chronic respiratory symptoms is made.
- Recommend annual reevaluation for respirator use.

Education

1. Emphasize importance of timely diagnostic evaluation for persistent respiratory symptoms of unknown cause.
2. Emphasize importance of timely assessment for prophylaxis after PPD test conversion.
3. When appropriate, emphasize importance of compliance with prophylaxis or treatment.
4. Encourage reporting of new or changed respiratory symptoms.
5. Report respirator intolerance.

Potential Accommodations

Allow temporary restrictions from respirator use pending completion of diagnostic evaluation and initiation of treatment.

Miscellaneous Conditions. Pneumothorax (question 3h):

1. Individuals with a history of an isolated spontaneous pneumothorax more than 3 years ago or an episode between 1 and 3 years ago but associated with a known cause (such as iatrogenic or traumatic) without sequelae:
   - Approve for respirator use without additional consultation of evaluation
   - For SCBA or work in IDLH atmospheres, defer respirator approval pending receipt of documentation from treating clinician.
   - Provide employee and employer with an appropriate timeline for completion of the evaluation and review for approval for respirator use.
   - On receipt of documentation consistent with this clinical scenario, approve for all respirators without additional consultation or evaluation. (A clinical examination is recommended prior to SCBA approval.)
   - Recommend reevaluation by brief questionnaire (see Table 3-3) or age-related guideline (see Table 3-4), barring any change in health status or working conditions.
2. Individuals with recurrent spontaneous episodes or an episode within the last 3 years:
   - Defer respirator approval pending advanced medical evaluation and optimization of therapeutic management.
   - Provide employee and employer with an appropriate timeline for completion of the evaluation and review for approval for respirator use.
   - Defer recommendation for periodic reevaluation on results of advanced medical evaluation.

Additional Considerations for Advanced Respirator Approval
Individual with Recurrent Spontaneous Episodes or an Episode within the Last 3 Years

- Consider whether the pneumothorax is primary or secondary to an underlying pulmonary disease. If secondary to an underlying pulmonary disease, defer respiratory approval pending advanced medical evaluation and to ensure optimal management. Apply relevant guideline for underlying disease as well.

- Provide employee and employer with an appropriate timeline for completion of the evaluation and review for approval for respirator use.
- Consider that chemical pleurodesis substantially decreases the risk of pneumothorax recurrence.
- If two or more episodes of pneumothorax have occurred within the last 3 years, refer to a pulmonary specialist for assessment and recommendations regarding both pleurodesis and respirator approval.
- If employee is at high risk of imminent recurrence, recommend accommodations and restrictions specific to the work site.
- Consider the hazards of the work environment and the consequences of respirator or job performance failure in making final approval decision.
- If the employee will work in an IDLH atmosphere or in a job that puts the public or coworker safety at risk, consider denying respirator approval. Otherwise, involve the individual in decision making regarding restrictions or denial of respirator approval.

Education

1. Review risk of recurrence of pneumothorax.
2. Note uncertainty of whether respirator use increases risk of recurrence but that recurrence while wearing a respirator may lead to additional medical sequelae because of respirator intolerance or inability to perform job tasks.
3. Emphasize strong relationship between smoking and pneumothorax, and encourage smoking cessation.
4. When appropriate, review usefulness of pleurodesis in reducing the risk of recurrence.
5. Report respirator intolerance.

Potential Accommodations

1. Allow temporary restrictions from respirator use pending completion of any further diagnostic evaluation and initiation of treatment.

2. Provide a positive pressure respirator to possibly decrease the risk that respirator use may theoretically incur.

3. Eliminate requirement for work in an IDLH atmosphere or where public and coworker safety is at risk.

4. Eliminate requirement for respirator use.

Lung Cancer (question 3i)

1. Individuals with a diagnosis of lung cancer in remission for more than 5 years and without pulmonary sequelae:
    - Approve for all respirators without additional consultation or evaluation. (A clinical examination is recommended prior to SCBA approval.)
    - Recommend annual reevaluations.
2. Individuals with a diagnosis of lung cancer within the last 5 years or with pulmonary sequelae despite a longer remission:
    - Defer respirator approval pending advanced medical evaluation.
    - Provide employee and employer with an appropriate timeline for completion of the evaluation and review for approval for respirator use.
    - Defer recommendation for periodic reevaluation on results of advanced medical evaluation.

Additional Considerations for Advanced Respirator Approval
Individuals with a diagnosis of lung cancer within the last 5 years or with pulmonary sequelae despite a longer remission:

- Obtain medical information from the treating physician about status, follow-up, active disease, and past and current treatment, as well as for documentation of any significant sequelae of treatment.
- Ensure that appropriate medical follow-up has occurred and base respirator approval on type of sequelae.
- Obtain recommendations of the treating physician about respirator use.
- Consider the hazards of the work environment and the consequences of respirator or job performance failure in making final approval decision.
- If employee's current pulmonary condition is likely to interfere with safe use of a respirator while performing assigned tasks, recommend accommodations and restrictions specific to the work site. If no accommodations or restriction will make safe use of a respirator possible, recommend denial of respirator approval.
- Based on the nature of the sequelae, current health status, and treatment, individualize recommendations for periodic reevaluations.

Education

1. Encourage smoking cessation when appropriate.
2. Encourage compliance with routine monitoring for disease activity if individual is in remission.
3. Encourage compliance with treatment recommendations if treatment is ongoing.
4. Report problems with respirator tolerance.

Potential Accommodations

1. Allow temporary restrictions from all respirator use pending completion of treatment, or schedule respirator use based on symptom fluctuation with treatment cycles.
2. Restrict from potential exposure to hazardous agents that may interact with treatment or are carcinogenic.

3. Eliminate requirement for respirator use.

Fractured Ribs (question 3j)

1. Individuals whose ribs were fractured more than 2 months ago without sequelae or underlying disease that may have lead to pathologic fracture:
   - Approve for respirator use without restrictions and without additional consultation. (A clinical examination is recommended prior to SCBA approval.)
   - Recommend reevaluation by brief questionnaire (see Table 3-3) or age-related guideline (see Table 3-4), barring any change in health status or working conditions.
2. Individuals whose ribs were fractured less than 2 months ago or who have sequelae:
   - Defer respirator approval pending advanced medical evaluation and optimization of therapeutic management.
   - Provide employee and employer with an appropriate timeline for completion of the evaluation and review for approval for respirator use.
   - Defer recommendation for periodic reevaluation on results of advanced medical evaluation.

Additional Considerations for Advanced Respirator Approval
Individuals whose ribs were fractured less than 2 months ago or with sequelae:

- Review medical records for evidence of underlying disease that may have lead to a pathologic fracture or result in significant sequelae.
- Pending resolution of pleuritic pain associated with uncomplicated fractures, defer respirator approval for job tasks that will incur increased ventilatory effort or relevant musculoskeletal stresses.
- If fractures are pathologic, deny respirator approval for job tasks that will incur increased ventilatory effort or relevant musculoskeletal stresses.

- If relevant sequelae persist more than 2 months after rib fracture, ensure appropriate medical follow-up and base respirator approval on type of sequelae.
- Recommend reevaluation by brief questionnaire (see Table 3-3) or age-related guideline (see Table 3-4), barring any change in health status or working conditions.
- If rib fractures are complicated by underlying disease, individualize recommendations for periodic reevaluations.

Education

1. Inform workers about usual time course for healing of uncomplicated rib fractures.
2. Tailor any additional education needed to associated sequelae or disease.

Potential Accommodations

1. Allow temporary restrictions from respirator use pending healing of rib fractures.
2. Permit respirators or job tasks that will not incur increased ventilatory effort or relevant musculoskeletal stresses.
3. Eliminate requirement for respirator use.

Thoracic Surgery or Injury Other than Fractured Ribs (question 3k)

1. Individuals with thoracic surgery or injury more than 1 year ago and without sequelae:
   - Unless associated with an underlying condition otherwise specified in these guidelines, approve for all respirators without additional consultation or evaluation. (A clinical examination is recommended prior to SCBA approval.)
   - Recommend reevaluation by brief questionnaire (see Table 3-3) or age-related guideline (see Table 3-4), barring any change in health status or working conditions.
2. Individuals with thoracic surgery or injury less than 1 year ago or sequelae from more remote traumatic or surgical events:
   - Defer respirator approval pending advanced medical evaluation.

- Provide employee and employer with an appropriate timeline for completion of the evaluation and review for approval for respirator use.
- Base recommendation for periodic reevaluation on results of advanced medical evaluation.

Additional Considerations for Advanced Respirator Approval
Individuals with thoracic surgery or injury less than 1 year ago or sequelae from more remote traumatic or surgical events:

- Review medical records for evidence about the type of chest injury or surgery and documentation of any significant sequelae or underlying disease.
- Ensure that appropriate medical follow-up has occurred and base respirator approval on the type of sequelae.
- Consider the hazards of the work environment and the consequences of respirator or job performance failure in making final approval decision.
- If the employee's current pulmonary condition is likely to interfere with safe use of a respirator while performing assigned tasks, recommend accommodations and restrictions specific to the work site. If no accommodations or restriction will make safe use of a respirator possible, recommend denial of respirator approval.
- Based on the nature of sequelae or any underlying disease, individualize recommendations for periodic reevaluations.

Education

1. Tailor education to the specifics of the condition.
2. Report respirator intolerance

Potential Accommodations

1. Allow temporary restrictions from respirator use pending resolution of sequelae associated with chest injury or surgery.
2. Permit respirators or job tasks that will not incur increased ventilatory effort or relevant musculoskeletal stresses.

3. Eliminate requirement for respirator use.

Cardiovascular Conditions or Symptoms

5. Have you had any of the following cardiovascular or heart problems (a–h)?

6. Have you ever had any of the following cardiovascular or heart symptoms (a–f)?

Questions 5 and 6 of the OSHA Respiratory Medical Evaluation Questionnaire address cardiovascular diagnoses and nonspecific symptoms often associated with cardiovascular disease. The many conditions and symptoms identified by these questions fall into the following more general categories that will be used to organize this section: ischemic heart disease (question 5a, c), heart failure (question 5d, e), cardiac arrhythmias (question 5f), cerebrovascular disease (question 5b), cardiovascular risk factors (question 5g), and symptoms without diagnosis (question 6a–f).

Examiners will encounter these common conditions frequently in the United States. About 21 million new cases of heart disease are reported annually, and it remains the leading cause of death.[55] There are almost 3 million stroke survivors, and almost 500,000 new strokes occur per year.[56] Another 50 million Americans have hypertension.[57] Although valvular disease, peripheral vascular disease, and some other cardiovascular conditions are also prevalent, they will rarely be disqualifying in the absence of current symptoms, arrhythmias, heart failure, or inability to perform similar tasks without a respirator.

The risk of sudden incapacitation or aggravation of a cardiovascular condition while workers wear a respirator may be reason for restricting their use of a respirator. Loss of consciousness or work capacity while wearing a respirator in an IDLH atmosphere, operating dangerous machinery, or working at heights may result in death or serious health consequences. As noted in Chapter 1, respirators, particularly SCBAs combined with other PPE, can substantially stress the cardiovascular system, increasing blood pressure, heart rate, and oxygen consumption. Physicians and LHCPs may appropriately approve most people with mild cardiovascular disease to wear respirators, but the presence of cardiovascular disease warrants careful attention to an individual's functional reserve, the specific aerobic demands of the assigned job tasks, and the consequences of incapacitation.

Agencies with an interest in public safety, such as the DOT, FAA, NRC, and National Fire Protection Administration (NFPA), specify disqualifying cardiovascular conditions that may lead to sudden incapacitation. Expert task forces have promulgated recommendations for qualifying athletes for sports that instruct PLHCPs performing evaluations for respirator use. Many workers wear respirators for work that is much less demanding or safety sensitive than that covered by these regulations and guidelines. Referral to the agencies and task force proscriptions, however, instructs the physician or LHCP evaluating individuals who will work in an IDLH atmosphere or in jobs where sudden incapacitation may result in significant adverse consequences.

In 49 CFR 391.41(b)(4), the DOT requires disqualification of any person who has a "... current clinical diagnosis of myocardial infarction, angina pectoris, coronary insufficiency, thrombosis, or any other cardiovascular disease of a variety known to be accompanied by syncope, dyspnea, collapse, or congestive cardiac failure...." The agency has specific rules regarding qualification of drivers with hypertension. In addition, the Federal Highway Administration's (FHWA) Conference on Cardiac Disorders and Commercial Drivers details recommendations for many specific conditions reviewed in the following sections.[58]

The Nuclear Energy Institute's Personnel Access Data System[6] considers a resting pulse rate greater than 100 or less than 40 beats per minute or blood pressure greater than 160/100 mmHg as potentially disqualifying conditions. The manual also directs PLHCPs to identify ischemic disease, stroke, valvular disorders, cardiac surgery, and cardiac enlargement. In all cases, the PLHCP may use clinical judgment in making final decisions about respirator qualification.

NFPA 1582[5] disqualifies firefighters with any of the following cardiac conditions: angina pectoris; left bundle-branch block or second-degree type II atrioventricular block; current heart failure; acute pericarditis, endocarditis, or myocarditis; recurrent syncope; and an automatic implantable cardiac defibrillator. The NFPA guidelines also consider several other cardiovascular conditions as potentially disqualifying, including significant valvular lesions of the heart, several arrhythmias, hypertrophy of the heart, chronic pericarditis, endocarditis, and myocarditis.

Not surprisingly, the FAA has the strictest requirements.[9] This agency prohibits qualification of first-, second-, and third-class airmen with a history or clinical diagnosis of any of the following conditions: myocardial infarction; angina pectoris; coronary heart disease that has required treatment or, if untreated, that has been symptomatic or clinically significant; cardiac valve replacement; permanent cardiac pacemaker implantation; or heart replacement.

Cardiovascular conditions associated with sudden death in athletes include hypertrophic cardiomyopathy, congenital heart disease, coronary artery anomalies, Marfan syndrome, the long QT syndrome, and arrhythmogenic right ventricular dysplasia.[59] The best estimate of deaths from these causes, however, suggests a frequency of only 1 death per 300,000 high school athletes.

Ischemic Heart Disease and Heart Failure

Exceeding the aerobic or pump capacity of an individual with heart disease may lead to catastrophic results not only for the worker but also for coworkers or the public. With careful attention to matching work capacity with job and respirator demands, many workers in class I or II of the New York Heart Association's (NHYA) functional classification of people with heart disease can work while wearing respirators[60] (Table 4-2). Modern medicine presents the clinician with several tools to broadly stratify the risk categories of workers with known coronary artery disease (CAD):

1. *High risk:* Left ventricular dysfunction at rest
2. *Moderate risk:* Normal resting left ventricular function but exercise- or stress-induced ischemia, arrhythmia, hypotension, or decreased exercise capacity
3. *Low risk:* All others.

Medical noncompliance, severe psychosocial job stress, or absence of symptoms with ischemia may increase the risk category.[61]

A number of scenarios related to ischemic heart disease present themselves, ranging from stable or unstable angina, to remote myocardial infarction (MI), to recent (< 3 months) MI, to a history of various cardiac interventions including coronary artery bypass grafting (CABG), percutaneous transluminal angioplasty (PTCA), and insertion of pacemakers or automatic implantable cardiac defibrillators.

Although a minority of deaths from CAD are sudden, most respirator workers with this condition will have enough warning before collapsing to summon help or remove themselves from hazardous environments. After an MI or CABG, the FHWA Conference on Cardiac Disorders and Commercial Drivers recommends a 3-month wait and a

Table 4-2. Cardiovascular Functional Classification of Patients with Cardiac Disease

Class	New York Heart Association functional description	Activity description
I	No symptoms or limitations with ordinary physical activity	Able to complete any activity requiring 7 METs[a] or less, e.g., can carry 24 lb up eight steps, carry objects that weigh 80 lb, do outdoor work (shovel snow, spade soil), do recreational activities (skiing, basketball, squash, handball, jog/walk 5 mph)
II	Slight activity limitation; comfortable at rest, but ordinary physical activity results in fatigue, palpitation, dyspnea, or anginal pain	Able to complete any activity requiring 5 METs or less, e.g. have sexual intercourse without stopping, garden, rake, weed, roller skate, dance fox trot, walk at 4 mph on level ground, but cannot and do not complete activities requiring 7 METs or more
III	Marked activity limitation; comfortable at rest, but less than ordinary physical activity results in fatigue, palpitation, dyspnea, or anginal pain	Able to complete any activity requiring 2 METs or less, e.g., shower without stopping, strip and make bed, clean windows, walk 2.5 mph, bowl, play golf, dress without stopping, but cannot and do not complete any activities requiring 5 METs or more
IV	Unable to carry out any physical activities without cardiovascular symptoms; symptomatic at rest	Unable to complete activities requiring 2 METs or more; cannot carry out any activities listed above

[a]METs = metabolic equivalents.

Source: Adapted from Braunwald E. The history. In: Braunwald E (ed.), *Heart Disease: A Textbook of Cardiovascular Medicine.* 5th ed. Philadelphia: WB

negative EST before returning a person to commercial driving. The conference required only a 3-week waiting period and a negative EST after a PTCA. Patients with stable angina and mild or moderate symptoms for the past 3 months can be qualified. Several organizations and some literature have recommended commercial driving disqualification of individuals using automatic implantable cardioverter-defibrillators (AICDs).[62] An individual may lose consciousness during the 20 seconds it may take for the AICD to discharge.

Interview, physical examination, electrocardiography, ambulatory ECG monitoring (AEM), ESTs, echocardiograms, Holter monitoring, and other advanced diagnostic methods provide the necessary armamentarium both to define the broad risk category and to focus the assessment on specific job demands. Evaluation for respirator use begins with an assessment of whether the individual can perform the assigned tasks without a respirator. When the concern rests with the additional load provided by a respiratory protection device, a physician may conduct a preplacement EST with the subject wearing the device to assess the additional cardiac demand it creates.

Using results of these examinations, a cardiologist can predict the risk of an adverse cardiac event and prescribe a workload that should not be exceeded. An individual's *peak exercise capacity* is defined as the maximal ability of the cardiovascular system to transport oxygen to exercising skeletal muscle and the ability of exercising muscle to extract oxygen from the blood. Maximal oxygen uptake (Vo_{2max}) provides the best measure of this exercise capacity. Because determination of Vo_{2max} is cumbersome, the aerobic accomplishment of an individual completing an EST is reported as multiples of resting oxygen consumption (METs). For example, an individual completing stage III of the Bruce protocol achieves between 8 and 10 METs. Before releasing an individual who has had an MI to unsupervised exercise, the ACSM requires a normal EST with exercise up to at least 8 METs.[63] Rough estimations of the energy costs in METs of various activities and job tasks have been published, but these do not incorporate psychological stress or the simultaneous use of a respirator.[61] To offer a frame of reference, shoveling at 10 cycles per minute of 14-lb loads is roughly equivalent to 8 to 9 METs (see Table 2-3).

Cardiac Arrhythmias

The Bethesda Conference of Specialists in Sports Medicine identified arrhythmias that pose a significant risk of sudden incapacitation.[64] These conditions are relevant to individuals wearing respirators whose work will create substantive aerobic demands. The arrhythmias of biggest concern catalogued by the conference are prolonged QT syndrome, ventricular tachyarrhythmias, exercise-induced superventricular tachycardia, atrial fibrillation, and atrial flutter with accelerated ventricular response during exercise. The conference allowed exercise with sinus node dysfunction as long as bradycardia increased appropriately with activity but disallowed sports if syncope or near-syncope occurred. Individuals with a history of impaired consciousness or fatigue secondary to bradycardia are returned to exercise after treatment when asymptomatic for 3 to 6 months. The conference permits patients treated for symptomatic tachycardia/bradycardia syndrome to engage in low-intensity exercise after a 3- to 6-month symptom-free period. Pacemakers exclude only those activities with a potential for bodily collision. Prolonged QT syndrome requires restriction from all sports. Atrial fibrillation or flutter should limit activities only if associated with structural heart disease or abnormal heart rates during exercise. Similarly, in the absence of structural heart disease, the Bethesda Conference allows exercise for asymptomatic patients with bundle-branch block, atrioventricular (AV) block, ventricular tachycardia, premature ventricular contractions, and superventricular tachycardia. Successful treatment of symptomatic patients usually should be followed by a 3- to 6-month monitoring period for symptoms before return to physically stressful activity. Structural heart disease generally requires limitation to low-intensity exercise.

Stroke

Stroke raises two concerns for the respirator user. First, do sequelae from a past stroke impair the worker's ability to perform a job safely while wearing a respirator? The answer to this question depends entirely on the severity of the neurologic impairment. Any impairment may readily be ascertained on physical examination, functional capacity evaluation, or a work trial. Of greater complexity is the second concern: What is the likelihood of recurrence? The sudden onset of neurologic symptoms that characterize stroke can result in dangerous, even lethal, incapacitation of a worker wearing a respirator in a hazardous environment.

Strokes recur frequently. The greatest risk occurs during the first 30 days, with rates ranging from 3 to 10 percent. Different studies document long-term stroke recurrence rates ranging from 4 to 14 percent per year.[56] Although risk factors have been clearly elucidated for first stroke, predictors of recurrence remain uncertain.

Hypertension, Cardiovascular Risk Factors, and Other Asymptomatic Conditions

In the absence of limiting symptoms, limited aerobic capacity, or heart failure, few cardiovascular conditions warrant restrictions from respirator use. Notable exceptions include hypertension and those asymptomatic conditions associated with sudden cardiac death during heavy exertion: myocarditis, hypertrophic cardiomyopathy with or without obstruction, and valvular aortic stenosis.[65]

Physicians and LHCPs commonly evaluate asymptomatic workers with cardiovascular risk factors. Increases in blood pressure while wearing a respirator may precipitate an adverse cardiovascular event in susceptible individuals. In the absence of cardiovascular symptoms, measurable dysfunction, or high-grade hypertension, however, the risk of an adverse event while wearing a respirator is usually remote. Most important in the evaluation will be the number of risk factors in relationship to the job and respirator demands. As an illustration, the chance of a 40-year-old male with moderate hypertension alone developing ischemic heart disease over an 8-year period is about 5 percent, but it rises to 70 percent if all coronary risk factors are present.[66] Stage IV hypertension (210/120 mmHg) frequently leads to imminent catastrophic events and requires emergent medical attention before returning to work. Stage III hypertension(180–209/110–119 mmHg) is also associated with the potential for near-term adverse cardiovascular events and demands urgent medical care and follow-up. Respirator use by workers with stage III and IV hypertension should be deferred pending improved blood pressure control.

Using the risk of sudden incapacitation as justification, the DOT requires persons with hypertension to have their disease controlled to levels less than 160/90 mmHg to maintain their medical certification to drive a commercial vehicle.[8] The NEI disqualifies workers from wearing respirators in the nuclear power industry with hypertension over 160/100 mmHg unless, based on the specific clinical context, the physician or LHCP waives this requirement.[6] NFPA 1582[5] expects disqualification of firefighters with blood pressures more than 180/100 mmHg when the physician concludes that the hypertension is of sufficient severity to pose a significant risk to the safety and health of the candidate or others. Rather than selecting a disqualifying blood pressure lower than 180/110 mmHg, we recommend that physicians and LHCPs use ACSM's criteria of two or more risk factors (see Table 3-7) as grounds for deferring respiratory approval pending further evaluation.

Because of expense and problems with false-positive tests, we recommend that examiners restrict advanced cardiac evaluation of asymptomatic workers with risk factors to those who will wear SCBAs or, rarely, to those wearing other respirators engaged in highly physically demanding job tasks. Physicians and LHCPs should predicate their decision to perform an EST on an individual's burden of risk factors, as noted in Table 3-7.

Cardiovascular Findings or Symptoms without Diagnosis

Workers presenting with undiagnosed significant cardiovascular findings or symptoms require appropriate evaluation and treatment before completing the respiratory qualification process. Physicians and LHCPs frequently may identify asymptomatic cardiac murmurs on physical examination. The examination itself usually will permit diagnosis of the type of valvular heart disease, but estimation of its severity in the absence of symptoms typically requires an ECG, a chest x-ray, and Doppler echocardiography at a minimum. The determinants of sudden death and the development of symptoms with exercise in people with valvular heart disease are well understood, but little information predicts the rate of progression of the disease.[67] When significant valvular disease coexists with other cardiovascular abnormalities, stringent restrictions may be justified for people whose jobs and respirators place heavy demands on them. With the help of a cardiologist, physicians may substantially liberalize the recommendations with specific attention to the individual circumstances.

Interview

The physician or LHCP should interview the employee to gather additional information about the cardiovascular condition or symptoms (see Figure 4-1) and procure relevant medical records from the individual's treating practitioner. When obtained by accepted screening techniques, unexpectedly high blood pressure requires careful confirmation.[66] If the worker has cardiovascular symptoms or physical diagnostic findings but no diagnosis or therapeutic plan, the PLHCP must refer the individual to an appropriate clinician. Many individuals with cardiovascular symptoms are undertreated. Referral for more intensive treatment may resolve any concerns regarding use of a respirator.

Despite the wide variety of potential conditions, the following single set of questions serves to outline the critical clinical data needed to proceed to the assessment phase:

- Date of symptom onset and diagnosis
- Episodic or chronic symptoms
- Frequency of episodes over 12 months, if episodic, with date of last episode
- Current symptoms
- Factors that trigger symptoms
- Results and dates of diagnostic tests
- Medications and other treatments including hospitalizations
- Functional impairment, including existing occupational restrictions, history of lost work due to cardiovascular diagnoses or symptoms, impact on exercise or other strenuous activities, impact on activities of daily living
- History of respirator use and tolerance

Assessment

With few exceptions, workers who report cardiovascular disease or symptoms should be referred to an appropriate advanced clinician for further evaluation. The sensitivity of the questionnaire, however, may lead to positive responses that have no current clinical significance. In these circumstances, the physician's or LHCP's interview and medical record review alone are usually sufficient. Examples of responses clarified by interview that should not affect respirator approval include

- Transient pedal edema in the past
- Minor peripheral circulatory conditions
- Rare palpitations lasting a few seconds without other symptoms
- Fleeting chest pain
- Past cardiac symptoms evaluated and found not to originate from the cardiovascular system
- Well-controlled stage I or II hypertension, without target organ disease, followed by a physician

The examiner cannot rely on symptoms alone as a gauge of a cardiac patient's ability to tolerate exertion while wearing a respirator. Silent ischemia occurs commonly, and arrhythmias may occur unexpectedly. EST, echocardiograms, AEM, and other advanced cardiac testing provide probabilistic guidance but not a crystal ball about what an individual with cardiac disease can tolerate. Because respirators can add substantially to the cardiac burden, tolerance of job tasks without a respirator does not guarantee that an individual can work safely with one. More confidence can be placed on EST or AEM performed while wearing the respiratory protection device and other necessary PPE. Because of the risk of death, however, work trials should not be conducted in the field. All simulated work testing must be done under the supervision of a physician.

Workers with some cardiac diseases may be particularly susceptible to the toxic effects of certain agents. Methylene chloride, carbon monoxide, and oxygen-deficient atmospheres may, for example, compound ischemia. Some substances, such as halogenated hydrocarbons, may precipitate arrhythmias. Placement of people with cardiac disease in settings with the expectation that a respirator provides sufficient protection from cardiotoxic hazards may lead to catastrophe.

An affirmative answer to any of the following questions may indicate the need for respirator denial or restrictions and should prompt advanced medical evaluation:

- Are the cardiovascular symptoms sufficient to place physiologic limitations on the person's ability to perform activities similar to those which will be required while wearing a respirator?
- Are the psychophysiologic job and respirator demands likely to exceed the worker's measured cardiac capacity of event-free exertion?
- Are the cardiovascular symptoms likely to provoke noncompliance with respirator use?
- Does the individual have uncontrolled stage III or IV hypertension or milder grades of hypertension with multiple risk factors?
- Is the person at high, imminent risk of an adverse cardiovascular event?
- Does the person use medications likely to affect psychomotor functions (question 7a)?
- Does the person have a history of respirator intolerance due to cardiovascular problems?
- Does the person have an episodic condition that could temporarily interfere with safe job performance while wearing a respirator?

- Will the person be working in a hazardous environment that could precipitate an adverse cardiovascular event in this individual if PPE failed?
- Will the person be working in a safety-sensitive position or an IDLH atmosphere?

Plan

Respirator Approval Decision

Physicians or LHCPs can find recommendations in this section for workers with cardiovascular disease or symptoms with reference to the following headings:

- General considerations for all cardiovascular conditions
- Symptoms or conditions irrelevant to respirator use
- Ischemic heart disease and congestive heart failure
- Cardiac arrhythmias and ECG abnormalities
- Stroke
- Hypertension and other cardiovascular risk factors
- Miscellaneous asymptomatic cardiac conditions

The physician or LHCP should consult the first segment about general considerations before referring to the guidelines specific to a particular worker's cardiovascular condition. Employees with a single cardiovascular condition may respond affirmatively to several components of questions 5 and 6. Responses to the component questions may relate to more than one condition. Therefore, we encourage the physician or LHCP to scan both diagnostic headings and question numbers for relevant guidance within this section on cardiovascular conditions.

General Considerations for all Cardiovascular Conditions (questions 5 and 6)

1. Individuals with cardiac disease from any cause resulting in marked limitation of physical activity (NHYA class III or IV; see Table 4-2):
    - Defer respirator approval pending advanced medical evaluation.

- Provide employee and employer with an appropriate timeline for completion of the evaluation and review for approval for respirator use.
- Review medical records. If questions remain about whether an individual is optimally medically managed, advanced clinicians should refer to the attending physician or cardiologist.
- If function improves, consider approval as outlined below; otherwise, they should recommend denial of respirator qualification.

2. Individuals with cardiac disease of any cause in NHYA class I or II (see Table 4-2) and demonstrated or probable functional capacity to perform job tasks similar to those which will be performed while wearing a respirator:
   - Defer respirator approval pending advanced medical evaluation.
   - Provide employee and employer with an appropriate timeline for completion of the evaluation and review for approval for respirator use.
   - Review medical records and ensure that appropriate medical follow-up has occurred within 3 months. If questions remain about whether an individual is optimally medically managed, advanced clinicians should refer to the attending physician or cardiologist. If function improves, they should consider approval as outlined below.
   - Obtain recommendations of the treating physician about respirator use while performing specific job tasks.
   - Consider whether additional advanced cardiac testing may be helpful in assessing the risk of an adverse event.
   - Customize recommendations for specific cardiac conditions by consulting the guidelines below.
   - Consider the hazards of the work environment and the consequences of respirator or job performance failure in making a final approval decision.
   - With the attending physician's written approval, approve the worker for respirator use, except for SCBA. Restrict from job

tasks likely to exceed the individual's cardiac reserve or potentially to lead to exposure to an environment that could precipitate a cardiac event.

- If the employee's current cardiac condition is likely to interfere with safe use of a respirator while performing assigned tasks, recommend accommodations and restrictions specific to the work site. If no accommodations or restriction will make safe use of a respirator possible, advanced clinicians should recommend denial of respirator approval.
- Individualize recommendations for periodic reevaluations based on risk factors, stability and severity of disease, and compliance with medical treatment.

Education

1. Encourage smoking cessation and reducing other modifiable risk factors when appropriate.
2. Encourage compliance with treatment recommendations.
3. Urge reporting of new or persistent symptoms or changes in diagnosis.
4. Review additional cardiovascular burden of respirator use.
5. Report for reevaluation for difficulty performing job or if respirator use precipitates symptoms.
6. Decline use of respirator for temporary exacerbations of cardiac condition.
7. Consider referral to a cardiac rehabilitation program to safely improve conditioning.
8. Suggest informing physician or LHCP or employer of any need for accommodations.
9. Encourage pneumovax and annual flu immunization.

Potential Accommodations

1. Provide a PAPR to decrease the work of breathing.
2. Allow temporary restrictions from respirator use during transient decreases in cardiac function.
3. Pending cardiac rehabilitation or treatment likely to improve function, provide work that does not require a respirator.

4. Eliminate job requirement to work on a hazardous response team, in an IDLH atmosphere, or in specific environments that contain triggers that would aggravate the cardiac disease if the respirator or other PPE failed.
5. Eliminate job requirement to work with a respirator.

The Job Accommodation Network lists many ideas to help accommodate workers with cardiac conditions.[68]

Symptoms or Conditions Irrelevant to Respirator Use (questions 5e–h, 6a–d)

1. Individuals reporting the following symptoms or conditions: (a) transient pedal edema in the past, (b) minor peripheral circulatory conditions such as varicose veins, cold hands and feet, or Raynaud's syndrome, (c) rare palpitations lasting a few seconds without other symptoms, (d) fleeting chest pain, (e) past cardiac symptoms evaluated and found not to originate from the cardiovascular system, (f) heart murmurs previously evaluated and found to be insignificant, (g) well-controlled stage I or II hypertension followed by a physician but without target-organ disease or sufficient risk factors to warrant an EST before exercise:
   - Approve for respirator use without restrictions and without additional consultation. (A clinical examination is recommended prior to SCBA approval.)
   - If worker is a well-controlled stage I or II hypertensive without target-organ disease or other risk factors but SCBA will be used, consider referral for possible EST; otherwise, approve without additional consultation.
   - Recommend annual reevaluations for workers with well-controlled stage I or II hypertension.
   - For other listed symptoms, PLHCPs should recommend reevaluation by brief questionnaire (see Table 3-3) or age-related guideline (see Table 3-4), barring any change in health status or working conditions.
2. Individuals with other cardiovascular conditions or undiagnosed symptoms:
   - Defer respirator approval pending advanced medical evaluation.

- Provide employee and employer with an appropriate timeline for completion of the evaluation and review for approval for respirator use.
- Base recommendation for periodic reevaluation on results of advanced medical evaluation.

Ischemic Heart Disease and Heart Failure (questions 5a, c–e, 6a–c).

Workers with positive responses to any of these questions or diagnoses should be referred to an appropriate physician for further medical evaluation (exclusive of insignificant problems, as listed earlier). Physicians or LHCPs should provide the employee and employer with an appropriate timeline for completion of the evaluation and review for approval for respirator use.

For individuals with a history of MI or cardiac surgery:

- Defer respirator approval pending waiting periods: (a) MI or CABG, more than 3 months; (b) PTCA, more than 3 weeks.
- Review medical records for evidence about current status, follow-up, past and current treatment, compliance, and documentation of any significant sequelae of disease or treatment.
- Obtain recommendations of the treating physician about respirator use.
- Consider whether additional advanced cardiac testing may be helpful in assessing risk. Per ACSM guidelines, after MI, obtain negative stress test with exercise up to at least 8 METs prior to approval. Require 10 METs for work with SCBA, thermal environmental extremes, or vigorous activity.
- Consider whether treatment that is more aggressive or supervised reconditioning in a cardiac rehabilitation program may resolve concerns about respirator approval.
- Consider the hazards of the work environment and the consequences of respirator or job performance failure in making a final approval decision.
- Approve the worker for respirator use but restrict him or her from job tasks likely to exceed his or her cardiac reserve or from work in an environment that could precipitate a cardiac event.

- If the employee's current cardiac condition is likely to interfere with safe use of a respirator while performing assigned tasks, recommend accommodations and restrictions specific to the work site. If no accommodations or restriction will make safe use of a respirator possible, advanced clinicians should recommend denial of respirator approval.
- Individualize recommendations for periodic reevaluations based on job tasks, risk factors, stability and severity of disease, and compliance with medical treatment.

Cardiac Arrhythmias and ECG Abnormalities (questions 5a, c, f–h, 6a–d).

Workers with positive responses to any of these questions or diagnoses should be referred to an appropriate physician for further medical evaluation (exclusive of insignificant problems, as listed earlier). Physicians or LHCPs should provide the employee and employer with an appropriate timeline for completion of the evaluation and review for approval for respirator use.

1. Individuals with prolonged QT syndrome:
   - Deny approval for respirator use in an IDLH atmosphere or for safety-sensitive work.
   - Restrict the worker to low-intensity exertion (<5 METs).
2. Individuals with sinus node dysfunction and a documented appropriate increase of heart rate with exertion. Approve for all respirators.
3. Individuals with sinus node dysfunction and a history of syncope, near-syncope, or unusual fatigue with exertion:
   - Restrict the worker to work effort below an intensity that produces symptoms.
   - Deny approval for respirator use in an IDLH atmosphere or for safety-sensitive work.
   - If treatment resolves condition, defer approval for respirator use in an IDLH atmosphere or for safety-sensitive work pending a 3- to 6-month asymptomatic period.
4. Individuals with atrial fibrillation or flutter, bundle-branch block, AV block, ventricular tachycardia, premature ventricular

contractions, or supraventricular tachycardia in the absence of structural heart disease, symptoms, or abnormal heart rates with exertion. Approve for all respirators and activity.

5. Individuals with atrial fibrillation or flutter, bundle-branch block, AV block, ventricular tachycardia, premature ventricular contractions, or supraventricular tachycardia in the presence of structural heart disease, symptoms, or abnormal heart rates with exertion:
   - Restrict the worker to work effort below an intensity that produces symptoms.
   - Deny approval for respirator use in an IDLH atmosphere or for safety-sensitive work.
   - If treatment resolves condition, advanced clinicians should defer respirator approval respirator use in an IDLH atmosphere or for safety-sensitive work pending a 3- to 6-month asymptomatic period.
6. Individuals with pacemakers that correct an underlying arrhythmia:
   - Restrict the worker from work that might result in bodily collision or exposure to intense electromagnetic fields (consult pacemaker manufacturer).
   - Approve for all respirators.
7. Individuals with automatic implantable cardioverter-defibrillators:
   - Restrict the worker from work that might result in bodily collision or exposure to intense electromagnetic fields (consult manufacturer).
   - Deny approval for respirator use in an IDLH atmosphere or for safety-sensitive work.

Stroke (question 6f).

Workers with positive responses to this question or with this diagnosis should be referred to an appropriate physician for further medical evaluation (exclusive of insignificant problems, as listed earlier). Physicians and LHCPs should provide the employee and employer with an appropriate timeline for completion of the evaluation and review for approval for respirator use.

For individuals with a history of transient ischemic attacks or stroke:

- Obtain consultation from a neurologist to individualize recommendations for respirator use and work restrictions based on sequelae, duration since last event, and risk of future stroke.
- If the employee's current cerebrovascular condition is likely to interfere with safe use of a respirator while performing assigned tasks, recommend accommodations and restrictions specific to the work site. If no accommodations or restrictions will make safe use of a respirator possible, recommend denial of respirator approval.
- Individualize recommendations for periodic reevaluations based on job tasks, risk factors, stability and severity of disease, and compliance with medical treatment.

Hypertension and Other Cardiovascular Risk Factors (question 5g, h)

1. Individuals with poorly controlled stage I or II hypertension, evidence of target-organ disease, or meeting ACSM criteria for sufficient number of risk factors to warrant an EST before exercise (see Table 3-7):
   - Defer respirator approval pending review of medical records for evidence of other cardiovascular risk factors, diagnostic testing, past and current treatment, compliance, and documentation of any significant sequelae of disease or treatment.
   - Provide employee and employer with an appropriate timeline for completion of the evaluation and review for approval for respirator use.
   - If there have been no basic end-organ status evaluations within the last 12 months or questions remain whether the individual is optimally medically managed, refer to the attending physician or cardiologist.
   - In the absence of end-organ disease, approve for all respirators without restrictions. However, in the presence of risk factors meeting ACSM guidelines for EST before exercise (see Table 3-7), conduct an EST if the individual requires use of a SCBA or will use a respirator for work requiring greater than 7 METs of exertion.

- In the presence of end-organ disease, refer to appropriate guideline for further guidance.

2. Individuals with uncontrolled stage III and IV hypertension:
   - Defer respirator approval and refer to the attending physician for further medical evaluation and treatment.
   - Provide the employee and employer with an appropriate timeline for completion of the evaluation and review for approval for respirator use.
   - When blood pressure is controlled successfully (at least <180/110 mmHg), consider limited respirator approval (no SCBA, no thermal stress, and no activity > 7 METs) pending an EST.
   - In the presence of two or more risk factors meeting ACSM guidelines for an EST before exercise, conduct an EST if the individual requires use of an SCBA or will use other respirators for work requiring greater than 7 METs.
   - In the presence of end-organ disease, refer to the appropriate guideline for further guidance.

Miscellaneous Asymptomatic Conditions (questions 5h, 6f).

Myocarditis, hypertrophic cardiomyopathy with or without obstruction, and acquired valvular heart disease:

- Defer respirator approval and refer to a cardiologist for further medical evaluation and recommendations regarding respirator use.
- Provide the employee and employer with an appropriate timeline for completion of the evaluation and review for approval for respirator use.
- Consult the detailed recommendations of the Bethesda Conference for guidance on specific conditions when work demands exceed about 5 METs.[67]

Medication

7. Do you take medication for any of the following problems (a–d)?

In some cases, workers may report the use of medications, signaling health problems that otherwise may not have been identified. Their identification should prompt further inquiry about the nature of these conditions. Once having recognized these diseases, the physician or LHCP should refer to the appropriate guideline for advice.

In addition, many medications can affect an individual's ability to perform a job safely. In contrast to conventional concerns, at least one study has found no association between use of antidepressants, anxiolytics, or narcotics with work-related injuries. On the other hand, the use of antihistamines, antibiotics, and diabetes medication elevated the risk of occupational injury.[69] Frequently, by treating an underlying medical problem, medicine improves work performance. Pharmaceuticals, both over-the counter and prescribed, also can interfere with work or, rarely, make the user more susceptible to hazardous physical and chemical exposures. Side effects, dosing schedules, and possible interaction with occupational hazards require review during preplacement, return-to-work, and performance-initiated examinations.[70] This section will discuss only those instances in which medication may specifically interfere with a worker's safe use of a respirator.

One concern rests with those medicines which have the potential of compounding some of the known major physiologic stresses of a respirator: cardiovascular burden, thermal load, and dermal irritation. None of these medicines contraindicates work with respirators, but their use should prompt physicians and LHCPs to inquire regarding potential side effects, to consider a work trial, and to pay careful attention to respirator tolerance after approval. Some cardiovascular agents, such as beta blockers, may blunt an appropriate cardiovascular response to the increased demands of certain respirators. Performance may be impaired or, worse, orthostatic hypotension may result in syncope. Drugs that have anticholinergic side effects may affect a person's capacity to cope with the additional thermal demands of a respirator, potentially increasing the risk of heat-related illnesses. Dermal side effects of some medications may enhance the susceptibility of an individual to facial skin irritation from tight-fitting respirators.

Rarely, other problems may occur with medication use in respirator wearers. Although timely relief from respirator use allows an individual to maintain correct dosing schedules, respirators interfere with the urgent administration of medicines. Delaying the use of sublingual nitroglycerin or an inhaled bronchodilator may result in medical deterioration.

Interview

The physician or LHCP should interview the employee to gather additional information about the medicines used and why they are being used (see Figure 4-1). Inquiry also should be made about the use of over-the counter medications. In most cases, the essential data can be collected from the employee. Often, referral for further evaluation will not be necessary. Physicians and LHCPs should follow the relevant guidance for newly identified medical conditions. When the examinee cannot answer the following questions, the physician or LHCP should obtain a release to get the required information from the attending clinician:

- Name of medication(s), dosage, and route and schedule of administration
- Duration of use and of current dose
- Reason for use
- Effectiveness in treating condition
- Consequences of missing a dose
- Side effects, with particular attention to psychomotor effects, impairment of exercise, facial irritation, or decreased heat tolerance

Assessment

Problems that may occur for respirator users taking certain medications center on job demands, the work environment, and the duration of respirator use. An affirmative answer to any of the following questions may indicate the need for respirator denial or restrictions and should prompt advanced medical evaluation.

- Is a medication being used that signals an underlying medical condition that may reduce respirator tolerance or impair work performance?
- Will temporary delays in the use of a medication cause an adverse effect?
- Does the employee report side effects that may reduce respirator tolerance or impair work performance?

Plan

Respirator Approval Decision

1. Use of medications for conditions irrelevant to respirator use, without pertinent side effects, or need for unscheduled urgent use of medications:
   - Approve for respirator use without restrictions and without additional consultation. (A clinical examination is recommended prior to SCBA approval.)
   - Recommend reevaluation by brief questionnaire (see Table 3-3) or age-related guideline (see Table 3-4), barring any change in health status or working conditions.
2. Affirmative answers to assessment questions:
   - Defer respirator approval and consult with the attending physician to coordinate possible changes in the medication regimen to minimize problems with respirator use.
   - Base recommendation for periodic reevaluation on whether changes are made in medication that resolves concerns about respirator use.

Additional Considerations for Advanced Respirator Approval

1. Medication signals the presence of a medical condition relevant to respirator use:
   - Follow guidance relevant to that condition.
2. Urgent medication use may be necessary or medication is associated with side effects relevant to respirator use:
   - Defer respirator approval pending advanced medical evaluation for an alternative treatment regime that may make urgent use unnecessary and will not create undesirable side effects.
   - Provide the employee and employer with an appropriate timeline for completion of the evaluation and review for approval for respirator use.
   - Consider work trials to assess physiologic impairment associated with medication.
   - If alternative medication is not feasible and the medicine probably will interfere with safe use of a respirator while

performing assigned tasks, recommend accommodations and restrictions specific to the work site. If no accommodations or restriction will make safe use of a respirator possible, recommend denial of respirator approval.

- Consider the hazards of the work environment and the consequences of respirator or job performance failure in making the a final approval decision.
- Individualize recommendations for periodic reevaluations based on job tasks, reasons for use of medicine, and the potential for associated adverse effects while wearing a respirator.

Education

1. Encourage compliance with treatment recommendations.
2. Urge reporting of new symptoms or changes in medication.
3. Report for reevaluation for difficulty performing job or for respirator intolerance.
4. Urge the employee to refuse respirator use for temporary increased use of medication that creates side effects incompatible with safe respirator use.
5. Report changes in medications or side effects

Potential Accommodations

1. Allow temporary restrictions from respirator use during transient need for use of medicine that interferes with the safe use of a respirator.
2. Schedule respirator use to avoid conflicts with medication dosing schedule.
3. Pending changes in medication that will resolve concerns about respirator use, provide work that does not require a respirator.
4. Eliminate job requirement to work on a hazardous response team, in an IDLH atmosphere, or in specific environments where medication use may impair performance or lead to adverse effects.
5. Eliminate job requirement to work with a respirator.

History of Problems Wearing a Respirator

8. If you've used a respirator, have you ever had any of the following problems (a–e)?

A history of respirator intolerance provides the best predictive data for future intolerance and requires action by the physician or LCHP. By treating underlying medical or psychological conditions, providing alternative respirators, or offering other reasonable accommodations, physicians and LHCPs can assist many workers previously intolerant of respirators to wear one successfully. As noted in the introduction to this chapter, factors associated with an employer's respiratory protection program and not related to an individual's health status may lead to respirator noncompliance. Although an employer may not contract the LHCP to review the entirety of its program, the physician or LHCP has a responsibility to point out any recognized deficiencies that may affect respirator compliance, tolerance, and safe use.

Interview

The physician or LHCP should interview the employee to gather additional information about the respirator tolerance problem (see Figure 4-1). In many cases, physicians and LHCPs can easily identify solutions to the problem that will obviate the need for further evaluation. If simple solutions are not forthcoming, or when an underlying medical problem has caused the respirator tolerance problem , the physician or LHCP should refer the worker to an appropriate clinician. The physician or LHCP should collect the following data about problems associated with respirator use:

- Type of respirator worn at the time of the problem
- Frequency and severity of problem
- Relevant medical condition(s)
- Cause, if known, for problem, e.g., medical condition, poor respirator fit, other PPE, exposure to occupational hazard or environment, inadequacies of the employer's respirator protection program
- Impairment of job performance or use of respirator
- Solution, if any, to the problem

Assessment

Improved respirator fit, alternative respirators, medical treatment, work practice changes, or simple administrative measures frequently will improve respirator tolerance. In other cases, the history of intolerance provides the strongest possible evidence for respirator restriction or denial.

Plan

Respirator Approval Decision

1. The cause and solution to the problem have been identified previously:
   - Without additional consultation, approve for respirator use with restriction or accommodation as appropriate. (A clinical examination is recommended prior to SCBA approval.)
   - Recommend reevaluation by brief questionnaire (see Table 3-3) or age-related guideline (see Table 3-4), barring any change in health status or working conditions.
2. The cause or solution to the problem has not been identified, but the problem was one of comfort:
   - Recommend improved respirator fit and a work trial to assess tolerance with the alternative respirator.
   - Assuming satisfactory tolerance, approve for respirator use without restrictions and without additional consultation. (A clinical examination is recommended prior to SCBA approval.)
   - Recommend reevaluation by brief questionnaire (see Table 3-3) or age-related guideline (see Table 3-4), barring any change in health status or working conditions.
3. The work trial is not tolerated or the previous respirator intolerance was related to a significant psychophysiologic problem:
   - Defer respirator approval, obtain relevant documentation from the treating clinician, and refer the worker for advanced medical evaluation.
   - Provide the employee and employer with an appropriate timeline for completion of the evaluation and review for approval for respirator use.

- Base recommendation for periodic reevaluation on results of advanced medical evaluation.

Additional Considerations for Advanced Respirator Approval

1. Respirator intolerance signals the presence of a medical condition relevant to respirator use:
   - Follow guidance relevant to that condition.
2. Respirator intolerance is potentially related to an undiagnosed, untreated, or undertreated medical or psychological problem:
   - Defer respirator approval pending advanced medical evaluation and optimal treatment.
   - Provide employee and employer with an appropriate timeline for completion of the evaluation and review for approval for respirator use.
   - If medical correction of intolerance is not feasible and intolerance is likely to interfere with the safe use of a respirator while performing assigned tasks, recommend accommodations and restrictions specific to the work site. If no accommodations or restriction will make safe use of a respirator possible, recommend denial of respirator approval.
   - Consider the hazards of the work environment and the consequences of respirator or job performance failure in making final approval decision.
   - Individualize recommendations for periodic reevaluations based on job tasks, reasons for respirator intolerance, and potential for recurrence of intolerance.
3. Respirator intolerance is potentially related to problems with the employer's respiratory protection program:
   - Defer respirator approval pending identification and correction of the problems with the employer's respiratory protection program.

Education

1. Report problems with respirator tolerance.
2. Review key components of a respiratory protection program.

Potential Accommodations

1. Provide an alternative respirator or other PPE that is better tolerated.
2. Allow temporary restrictions from respirator use during transient health problems that interfere with respirator tolerance or correction of problems with the respiratory protection program.
3. Eliminate job requirement to work with a respirator.

Desire to Talk with a Health Care Professional

9. Would you like to talk to the health care professional who will review this questionnaire . . . ?

Question 9 of the OSHA questionnaire offers the respondent the opportunity to talk with the physician or LHCP who reviews the form. Concerns regarding confidentiality or confusion about how to answer a question may impede a worker from fully disclosing relevant health problems in writing. In some cases, the employee may have a medical problem entirely unrelated to respirator use but that deserves referral to an appropriate health care professional for follow-up. An affirmative response to question 9 does not mandate a medical examination unless the interview identifies a relevant condition that requires further evaluation.

Additional questions mandatory only for employees who have been selected to use a full-facepiece respirator or a self-contained breathing apparatus (SCBA) (questions 10–15).

Visual Problems

10. Have you ever lost vision in either eye (temporarily or permanently)?

11. Do you currently have any of the following vision problems?

Vision figures prominently in safe, efficient, and comfortable performance of most jobs. Sensory isolation and decrements in visual acuity and peripheral vision caused by mask fogging and visual obstruction compound a worker's own visual problems.[71] Further, the need for a tight facial seal necessitates the use of special mask glasses that can be uncomfortable to use and further diminish peripheral vision.

Research has demonstrated that wearing contacts does not increase the risk of ocular chemical injury in industrial environments.[72] In 1988, OSHA reversed its earlier dictate that workers could not wear contact lenses with full-face respirators.[73] Contacts therefore can be worn in hazardous environments and may be an alternative preferable to mask glasses. Unfortunately, however, supplied air dries contact lenses, making them intolerable for some workers. Color blind workers cannot depend on color to select appropriate cartridges. ELSI's do not depend on color vision. Salazar et al.[1] interviewed workers who indicated that visual problems seriously impaired their work performance and increased job risks. Others noted that visual effects were the most difficult aspect of wearing a respirator.

Specific guidelines for visual acuity, color vision, stereopsis, and visual field requirements pertain primarily to the type of work performed while wearing the respirator. NEI recommends at least corrected monocular 20/40 vision or accommodations that will otherwise allow the individual to accomplish the work tasks.[6]

Interview

The physician or LHCP should interview the employee to gather additional information about the visual problem (see Figure 4-1). In many cases, physicians or LHCPs can easily identify solutions to the problem that will obviate the need for further evaluation. If simple solutions are not forthcoming, or when the visual problem correlates with a medical problem that may affect respirator tolerance and safe use, the physician or LHCP should refer the worker to an advanced clinician. Several diseases or conditions can result in transient blindness, permanent blindness, or other visual disturbances including transient ischemic attacks, migraine, retinal detachment, multiple sclerosis, temporal arteritis, cerebrovascular accidents, diabetes, trauma, and others.

The physician or LHCP should procure relevant medical records from the individual's treating practitioner. If the worker has symptoms but no diagnosis or therapeutic plan, the physician or LHCP must refer the individual to an appropriate clinician. The physician or LHCP should collect the following data about visual problems that may impair respirator use:

1. Episodes of transient blindness in either eye:
   - Diagnosis
   - Triggers
   - Prodrome
   - Frequency, severity, duration, and date of last episode

- Prognosis
- Treatment

2. Permanent blindness in either eye:
   - Etiology
   - Treatment
3. Diminished visual acuity:
   - Any associated medical problem
   - Treatment
   - Correction with glasses or contact lenses
4. Color blindness:
   - Type and severity of color blindness
5. Other visual problems:
   - Diagnosis
   - Factors that aggravate or precipitate
   - Frequency, severity, and duration
   - Prognosis
   - Treatment

Assessment

Identification of the underlying medical problems associated with transient or permanent blindness usually warrants review with an appropriate clinician. Conditions causing transient blindness often can be treated successfully or have prodromes that allow time for safe escape from a hazardous environment. The cause, frequency, severity, duration, and date of the last episode provide further data to assess the risk of respirator use. Information about current functioning with a permanently blind eye may suffice for respirator decision making. When uncertainty about work capacity remains, work trials focused on job performance provide the needed additional information. Physicians and LHCPs should refer to the relevant protocols for guidance about associated medical problems.

Alternative respirators, changes in the approach to visual correction, or other accommodations usually can ameliorate visual problems.

An affirmative answer to any of the following questions may indicate the need for respirator denial or restrictions and should prompt an advanced medical evaluation.

- Is the visual problem associated with an underlying medical condition that may reduce respirator tolerance or impair work performance?
- Is the cause of the visual problem unknown?

Plan

Respirator Approval Decision

1. The cause and solution for the visual problem have been identified previously:
   - Defer respirator approval pending advanced medical evaluation if a significant underlying medical problem exists that may affect respirator use.
   - Provide the employee and employer with an appropriate timeline for completion of the evaluation and review for approval for respirator use.
   - In the absence of a significant medical condition, approve for respirator use without additional consultation, encouraging compliance with medical recommendations. (A clinical examination is recommended prior to SCBA approval.)
   - Specify restrictions or accommodations when relevant.
   - Recommend reevaluation by brief questionnaire (see Table 3-3) or age-related guideline (see Table 3-4), barring any change in health status or working conditions.
2. The cause and solution for the visual problem has not been identified previously:
   - Defer respirator approval pending advanced medical evaluation.
   - Provide the employee and employer with an appropriate timeline for completion of the evaluation and review for approval for respirator use.
   - Base recommendation for periodic reevaluation on results of advanced medical evaluation.

3. The visual problem was one of comfort or fit of mask, glasses, or contacts:

    - Recommend a trial of alternative respirators or visual corrective devices.
    - Assuming satisfactory tolerance, approve without additional consultation. (A clinical examination is recommended prior to SCBA approval.)
    - If satisfactory comfort and fit cannot be achieved, refer for advanced medical evaluation.
    - Recommend reevaluation by brief questionnaire (see Table 3-3) or age-related guideline (see Table 3-4), barring any change in health status or working conditions.

4. The worker reports color blindness:

    - When relevant, defer respirator approval and recommend a field test of the ability to correctly select color-coded respirator components or perform other job-critical color discriminating tasks.
    - If necessary, recommend accommodations to assist in the selection of color-coded respirator components.
    - If color blindness is irrelevant to the job or the worker's ability to perform the field test adequately, approve without additional consultation. (A clinical examination is recommended prior to SCBA approval.)
    - Recommend reevaluation by brief questionnaire (see Table 3-3) or age-related guideline (see Table 3-4), barring any change in health status or working conditions.

Additional Considerations for Advanced Respirator Approval
For an undiagnosed visual problem that continues to interfere with functioning or is associated with an underlying medical condition relevant to respirator use:

- Defer respirator approval pending advanced medical evaluation and treatment.
- Provide the employee and employer with an appropriate timeline for completion of the evaluation and review for approval for respirator use.

- Consult the relevant protocol for guidance about underlying medical problems.
- Consider work trials to assess respirator tolerance and capacity to perform the job safely.
- If visual problems interfere with the safe use of a respirator while performing assigned tasks, recommend accommodations and restrictions specific to the work site. If no accommodations or restriction will make performance of assigned tasks and safe use of a respirator possible, recommend denial of respirator approval.
- Consider the hazards of the work environment and the consequences of respirator noncompliance or job performance failure in making a final approval decision.
- Individualize recommendations for periodic reevaluations based on job tasks, reasons for visual problems, and potential for associated adverse effects while wearing a respirator.

Education

1. Report problems with respirator use or job performance.
2. Report worsening of visual problems.
3. If wearing contacts, encourage the individual to maintain a backup pair of contacts or mask glasses at work.

Potential Accommodations for Workers with Visual Problems

1. Provide an alternative respirator that is better tolerated.
2. Allow temporary restrictions from respirator use during transient health problems that interfere with vision.
3. Eliminate job requirement to work with a respirator.

The Web site Low Vision Gateway provides numerous ideas about accommodation visual problems.[74]

Auditory and Communication Disorders

12. Have you ever had an injury to your ears, including a broken eardrum?

13. Do you currently have any of the following hearing problems?

Respirators can aggravate communication problems caused by decreased auditory acuity or speech disorders. Respiratory devices that create noise and cover the ears further reduce a hearing-impaired individual's ability to hear language and important environmental sounds. Covering the mouth with a mask similarly may compound expressive language disorders. In addition to permanent and degenerative conditions, several diseases or physiologic states can result in transient hearing loss such as infection, noise exposure, or impacted cerumen. The sensory isolation inherent to respirators compounds underlying communication difficulties and may interfere with respirator tolerance, job performance, and safety.

Trauma, certain otologic diseases, or surgical interventions in the ear can produce a direct pathway between the external auditory canal and the respiratory system via the eustachian tube. Historically, OSHA has considered a perforated tympanic membrane a contraindication for respirator use. This dictum was based on the assumption that hazardous substances could migrate through the tympanic membrane. OSHA has changed its compliance directives because scientific analysis indicates that exposure would be minimal and rarely of clinical importance.[75] Despite this finding, conservative practitioners may wish to recommend positive-pressure respirators to ensure adequate protection in highly toxic or unknown atmospheres.[51]

Interview

The physician or LHCP should interview the employee to gather additional information about the auditory or communication problem (see Figure 4-1). The type of respirator and the job demands critically determine how important these problems may be. After reviewing relevant medical records from the individual's treating practitioner, physicians or LHCPs may easily identify solutions to the problem that will obviate the need for further evaluation. If simple solutions are not forthcoming, or when the sensory or expressive problem correlates with an underlying medical problem relevant to respirator tolerance and safe use, the LHCP should refer the worker to an advanced clinician. If the worker has symptoms but no diagnosis or therapeutic plan, the LHCP also must refer the individual to an advanced clinician.

The physician or LHCP should collect the following data about auditory or communication problems that may impair respirator use:

1. Episodes of transient hearing problems:
   - Diagnosis
   - Prodrome
   - Frequency, severity, duration, and date of last episode
   - Prognosis
   - Treatment
2. Permanent hearing loss:
   - Etiology
   - Severity
   - Treatment
   - Corrective devices
3. Other auditory or communication problems:
   - Diagnosis
   - Severity
   - Prognosis
   - Treatment

Assessment

Physicians and LHCPs rarely will need to deny respirator approval to employees with auditory or expressive language difficulties. Accommodations usually should be possible. The physician or LHCP should begin the assessment with a review of the importance of oral communication and hearing in the work environment. Identification of underlying, undertreated, or undiagnosed problems associated with transient or permanent hearing loss or communication difficulties warrants review with an appropriate clinician. Physicians or LHCPs should refer to the relevant protocols for guidance about associated medical problems. Treatment usually can ameliorate transient hearing loss, and assistive devices can improve hearing acuity. Alternative respirators may lessen the communication impairment and obviate even theoretical concerns about ruptured tympanic membranes. When uncertainty about work capacity remains, work trials can test an individual's ability to respond to specific critical communications at the work site.

An affirmative answer to any of the following questions may indicate the need for respirator denial or restrictions and should prompt an advanced medical evaluation.

- Is the auditory or communication problem associated with an underlying medical condition that may reduce respirator tolerance or impair work performance?
- Is the cause of the auditory or communication problem unknown?
- Has the auditory or communication problem previously impaired work performance or safety with or without respirator use?
- Does the employee have a perforated tympanic membrane (as a result of either disease or surgery)?

Plan

Respirator Approval Decision

1. All assessment questions are answered negatively, and hearing or oral communication are irrelevant to the job tasks:
   - Approve for respirator use without restrictions and without additional consultation. (A clinical examination is recommended prior to SCBA approval.)
   - Recommend reevaluation by brief questionnaire (see Table 3-3) or age-related guideline (see Table 3-4), barring any change in health status or working conditions.
2. All assessment questions are answered negatively, but hearing or oral communication are relevant to the job tasks:
   - Ensure that employee has been evaluated recently and provided with the most effective assistive device.
   - Defer respirator approval pending a work-trial assessment of response to critical communications.
   - If the employee performs adequately during the work trial, approve without additional consultation. (A clinical examination is recommended prior to SCBA approval.)

- If the employee has difficulty performing satisfactorily during the work trial, attempt to identify alternative respirators or communication methods that will correct the deficiency.
- If no satisfactory accommodation can be identified, refer for advanced medical evaluation.
- Recommend reevaluation by brief questionnaire (see Table 3-3) or age-related guideline (see Table 3-4), barring any change in health status or working conditions.

3. One or more of assessment questions have been answered affirmatively:
    - Defer respirator approval pending advanced medical evaluation and recommendations.
    - Provide the employee and employer with an appropriate timeline for completion of the evaluation and review for approval for respirator use.
    - Base recommendation for periodic reevaluation on results of advanced medical evaluation.

Additional Considerations for Advanced Respirator Approval

1. The auditory or communication problem is undiagnosed, continues to interfere with functioning, or is associated with an underlying medical condition relevant to respirator use:
    - Defer respirator approval pending advanced medical evaluation and recommendations.
    - Provide the employee and employer with an appropriate timeline for completion of the evaluation and review for approval for respirator use.
    - Consult the relevant protocol for guidance about associated medical problems.
    - Consider a work trial to assess respirator tolerance and capacity to perform the job safely.
    - If auditory or communication problems interfere with safe use of a respirator while performing assigned tasks, recommend accommodations and restrictions specific to the work site. If no accommodations or restriction will make performance of

assigned tasks and safe use of a respirator possible, recommend denial of respirator approval.

- Consider the hazards of the work environment and the consequences of respirator noncompliance or job performance failure in making a final approval decision.
- Individualize recommendations for periodic reevaluations based on job tasks, reasons for auditory and communication problems, and potential for associated adverse effects while wearing a respirator.

2. The employee has a perforated tympanic membrane:
   - If employee has never been evaluated for potential membrane repair, offer referral to an appropriate specialist.
   - In most cases, tympanic membrane perforation is irrelevant.
   - If the employee will work in an IDLH or unknown atmosphere, consider use of ear protection or recommend a positive-pressure respirator.
   - Recommend reevaluation by brief questionnaire (see Table 3-3) or age-related guideline (see Table 3-4), barring any change in health status or working conditions.

Education

1. Report problems with respirator use or job performance.
2. Report deterioration of auditory or communication conditions.

Potential Accommodations for Workers with Auditory and Communication Problems

1. Provide an alternative respirator that is better tolerated.
2. Allow temporary restrictions from respirator use during transient health problems that interfere with hearing or communication.
3. Eliminate the job requirement to work with a respirator.
4. Implement alternative communication methods such as hand signals, lights, vibrators, and others not dependent on hearing and oral language.

The Job Accommodation Network provides numerous ideas about the accommodation of individuals with hearing and communication problems.[76]

Musculoskeletal Problems

14. Have you ever had a back injury?

15. Do you currently have any of the following musculoskeletal problems (a–j)?

Use of a respiratory device requires the dexterity and strength to don and doff the equipment and to wear it while performing the required work duties. The individual also may change cartridges and maintain the respirator. The weight of an SCBA or a full mask may aggravate some workers' orthopedic problems. This heavy, awkward equipment may further limit a worker's ability to work or maintain his or her balance. Restrictions in cervical range of motion further limit peripheral vision already restricted by some respirators. The physician or LHCP rarely will encounter individuals with musculoskeletal difficulties that impair the ability to wear lighter and less bulky equipment but do not also limit the physical capacity to execute the assigned job tasks.

Interview

The physician or LHCP should interview the employee to gather additional information about the musculoskeletal problems (see Figure 4-1). Historically remote injuries or disorders that have resolved without sequelae rarely require further evaluation, especially if the individual has since performed challenging work without difficulty. Contemporary orthopedic complaints deserve in-depth inquiry and the procurement of relevant medical records and recommendations from the individual's treating practitioner. Unless these records clarify all concerns, the physician or LHCP should refer the worker to an appropriate clinician for physical examination. If the worker has symptoms but no diagnosis or therapeutic plan, the physician or LHCP also must refer the individual to an appropriate clinician.

Many methods exist for assessing the physical capacity of an individual with orthopedic conditions to wear a respirator while working. The examiner should conduct a physical examination to grossly screen for the capacity to grasp, pinch, squat, bend, twist, climb, and perform other essential tasks. Abnormalities or a history of joint unreliability requires specialized orthopedic evaluation. Besides historical evidence of sufficient abilities, the cheapest and most reliable method to assess the impact of questionable conditions involves work simulation. Physicians and LHCPs can easily explore concerns about a worker's dexterity and strength to doff and don a respirator, adjust its fit, and maintain it properly by observing the individual perform these tasks. Although not appropriate for emergency response or work in an IDLH atmosphere, brief work trials otherwise also can help assess any orthopedic limitations associated with wearing a respirator while working. When concerns remain for an individual with a more hazardous job description, a physical capacity evaluation in a medical setting can aid medical decision making.[77]

The physician or LHCP should collect the following data about musculoskeletal problems that may impair respirator use:

- Date of symptom onset and diagnosis
- Episodic or chronic symptoms
- Frequency of symptoms over the past 12 months if episodic, with date of last episode
- Current symptoms
- Factors that trigger symptoms
- Results and dates of diagnostic tests
- Medications and other treatments, including hospitalizations
- Functional impairment, including existing occupational restrictions, history of lost work due to orthopedic diagnoses or symptoms, impact on exercise or other strenuous activities, impact on activities of daily living
- History of respirator use and tolerance

Assessment

Only contemporary orthopedic problems cause problems wearing a respirator. The type of respirator and the job demands critically determine how limiting these musculoskeletal conditions may be. The physician or LHCP generally can provide respirator qualification to an individual whose work does not reproduce those physical stressors which precipitate symptoms. When uncertainty about work capacity remains, a work trial or functional capacity evaluation provides the needed additional information.

An affirmative answer to any of the following questions may indicate the need for respirator denial or restrictions and should prompt advanced medical evaluation.

- Are the musculoskeletal symptoms likely to provoke noncompliance with respirator use?
- Are the symptoms sufficient to place substantial physiologic limitations on the person's ability to perform activities similar to those which will be required while wearing a respirator?
- Does the person use medications likely to affect psychomotor functions (question 7)?
- Does the person have a history of respirator intolerance due to orthopedic problems (question 8)?
- Does the person have an episodic condition that could temporarily interfere with job performance or respirator tolerance while in an IDLH atmosphere or safety-sensitive position?
- Does the person have orthopedic symptoms sufficient to impair activity that have not been evaluated and treated?

Plan

Respirator Approval Decision

1. If the musculoskeletal problem occurred in the past and has no known sequelae:
   - Approve the worker for respirator use without restrictions or additional consultation.
   - Recommend reevaluation by brief questionnaire (see Table 3-3) or age-related guideline (see Table 3-4), barring any change in health status or working conditions.
2. Active musculoskeletal symptoms within the last year sufficient to impair activity or a positive response to one or more of the assessment questions:
   - Defer respirator approval pending procurement of relevant documentation from the treating clinician.
   - Refer for advanced medical evaluation if the job or respirator is likely to stress the affected area.

- Base recommendation for periodic reevaluation on results of advanced medical evaluation.

Additional Considerations for Advanced Respirator Approval

1. The orthopedic problem is undiagnosed and interferes with functioning relevant to job tasks and respirator use:
   - Defer respirator approval and conduct relevant further evaluation including orthopedic referral if necessary to ascertain the diagnosis and optimize treatment.
   - Provide the employee and employer with an appropriate timeline for completion of the evaluation and review for approval for respirator use.
2. The orthopedic problem may interfere with respirator use:
   - Defer respirator approval to optimize medical treatment by an appropriate advanced clinician.
   - Provide the employee and employer with an appropriate timeline for completion of the evaluation and review for approval for respirator use.
   - Consider work trials to assess respirator tolerance and capacity to perform the job safely.
   - Consider a physical capacity evaluation to predict ability to use a respirator if the job is safety-sensitive or involves work in an IDLH atmosphere.
   - If the orthopedic problems probably will be aggravated, interfere with job performance, or affect respirator tolerance, recommend accommodations and restrictions specific to the work site. If no accommodations or restrictions will make safe performance of the assigned tasks with a respirator possible, recommend denial of respirator approval.
   - Consider the hazards of the work environment and the consequences of respirator noncompliance or job performance failure in making a final approval decision.
   - Individualize recommendations for periodic reevaluations based on job tasks, the nature of the musculoskeletal problems, and the potential for associated adverse effects while wearing a respirator.

Education

1. Report problems with respirator use or job performance.
2. Report worsening of orthopedic problems.
3. Decline respirator use during transient flares of musculoskeletal conditions that affect respirator use.
4. Provide relevant biomechanical education to minimize the risk of work practices that aggravate the underlying condition.

Potential Accommodations for Workers with Musculoskeletal Problems

1. Provide an alternative respirator that is better tolerated.
2. Allow temporary restrictions from respirator use during transient orthopedic aggravations that interfere with respirator use.
3. Eliminate job requirement to work with a respirator.
4. Implement alternative work practices and job design to minimize musculoskeletal stressors.

The Job Accommodation Network provides numerous links to ideas about the accommodation of individuals with musculoskeletal problems.[78]

Summary

Either the OSHA questionnaire or an initial screening medical examination gathers most of the information necessary for physicians or LHCPs to make appropriate decisions regarding respirator approval. Depending on the surveyed population of workers, the sensitivity of the questionnaire frequently may precipitate follow-up by interview and additional information gathering from treating practitioners. More infrequently, the initial data should provoke referral to an appropriate advanced clinician for further evaluation and decision making. Professionalism, as well as the ADA, demands that whenever possible, physicians and LHCPs should base their medical counsel about respirator use on scientific evidence or at least medical consensus. Although little research has been conducted that provides the practitioner with confidence in the predictive power of any specific medical tests, a relevant scientific literature exists that at least supports the face validity of some data useful in decision making about respirator approval. In addition, several regulatory and nongovernmental organizations provide guidelines germane to respirator use qualification. Due to concerns regarding the predictiveness of medical evaluations for respirator use, medical evaluation programs should place more emphasis on assessing respirator users who report work or respirator tolerance problems. The health care provider's recommendations to employers should address whether a worker's disabilities impair performance of the essential functions of a job or present a direct health or safety threat to the worker or others. PLHCPs can assist workers and employers to identify reasonable accommodations that allow safe respirator use.

Medical evaluation for respirator use requires close cooperation between an employer and the contracted medical personnel. Physicians and LHCPs can only make rational decisions about respirator approval with a full understanding of a worker's assigned job tasks and the type of respirator to be used. Respirator approval for work in IDLH atmospheres, emergency response, or other jobs that affect public safety demands assessment that is more careful and conservative. Medical evaluation is only one factor in many that helps achieve the ultimate goal of appropriate respirator use. Moreover, respirator use is only one part of a comprehensive respiratory protection program.

Further Reading

Americans with Disabilities Act

Goldenthal N. *Understanding the Americans with Disabilities Act: A Compliance Guide for Health Professionals and Employees.* Beverly Farms, Mass.: OEM Press, 1993. *Full of forms, guides, and checklists, including the complete text of the ADA, this guide provides the information needed to interpret and implement this complex legislation.*

Resources for Rehabilitation. Resources for People with Disabilities and Chronic Conditions, 3d ed. Lexington, Mass.: Resources for Rehabilitation, 1996. A comprehensive resource guide that includes information on accommodations available for the most common disabling conditions, including spinal cord injury, low back pain, diabetes, multiple sclerosis, hearing and speech impairments, visual impairment, and epilepsy.

Relevant Guidelines and Federal Regulations

American National Standards Committee on Respiratory Protection, Z88. *American National Standard for Respiratory Protection: Respirator Use—Physical Qualifications for Personnel.* ANSI Z88.6-1984. New York: American National Standards Institute, 1984.

American Thoracic Society. Respiratory protection guidelines. *Am J Respir Crit Care Med* 1996; 154:1153–1165.

Technical Committee on Fire Service Occupational Medical and Health. *NFPA 1582 Standard on Medical Requirements for Fire Fighters,* 1997 edition. Quincy, Mass.: National Fire Protection Association, 1997.

The Nuclear Energy Institute Personnel Access Data System Project. *Determination of Medical Fitness to Use Respiratory Protective Equipment.* NEI 95-03. Washington: Nuclear Energy Institute, 1995.

American Nuclear Society Standards Committee Working Group ANS-3.4. *American National Standard for Medical Certification and Monitoring of Personnel Requiring Operator Licenses for Nuclear Power Plants.* ANSI/ANS 3.4 La Grange Park, Ill.: American Nuclear Society, 1996.

Medical Advisory Criteria for Evaluation Under 49 CFR Part 391.41. Available at *http://www.fmcsa.dot.gov/rulesregs/fmcsr/medical.htm.*

FAA Office of Aviation Medicine. *Civil Aeromedical Institute Aeromedical Certification. Part 67: Medical Standards and Certification.* Available at *http://www.cami.jccbi.gov/AAM-300/part67.html. These guidelines and regulations provide a 360-degree view of the current medical standards for approving workers for respirator use.*

References

1. Salazar MK, Takaro TK, Connon C, et al. A description of factors affecting hazardous waste workers' use of respiratory protective equipment. *Appl Occup Environ Hyg* 1999; 14:470–478.
2. American Thoracic Society. Respiratory protection guidelines. *Am J Respir Crit Care Med* 1996; 154:1153–1165.
3. Harber P, Merz B, Chi K. Decision model for optimizing respirator protection. *J Occup Environ Med* 1999; 41:356–365.
4. American National Standards Committee on Respiratory Protection, Z88. *American National Standard for Respiratory Protection: Respirator Use—Physical Qualifications for Personnel.* ANSI Z88.6-1984. New York: American National Standards Institute, 1984.
5. Technical Committee on Fire Service Occupational Medical and Health. *NFPA 1582 Standard on Medical Requirements for Fire Fighters,* 1997 ed. Quincy, Mass.: National Fire Protection Association, 1997.

6. The Nuclear Energy Institute Personnel Access Data System Project. *Determination of Medical Fitness to Use Respiratory Protective Equipment.* NEI 95-03. Washington: Nuclear Energy Institute, 1995.
7. American Nuclear Society Standards Committee Working Group ANS-3.4. *American National Standard for Medical Certification and Monitoring of Personnel Requiring Operator Licenses for Nuclear Power Plants.* ANSI/ANS 3.4 La Grange Park, Ill.: American Nuclear Society, 1996.
8. Medical Advisory Criteria for Evaluation under 49 CFR Part 391.41. Available at *http://mcregis.fhwa.dot.gov/medical.htmhttp://mcregis.fhwa.dot.gov/medical.htm.*
9. FAA Office of Aviation Medicine. *Civil Aeromedical Institute Aeromedical Certification. Part 67—Medical Standards and Certification.* Available at *http://www.cami.jccbi.gov/AAM-300/part67.html.*
10. OSHA Preambles: Respiratory Protection Standard. Available at *http://www.osha-slc.gov/Preamble/RP_html/RESPIRATORY3.html.*
11. Muhm JM. Medical surveillance for respirator users. *J Occup Environ Med* 1999; 41:989–994.
12. Occupational Medicine Forum. Guidelines for respirator users. *J Occup Med* 1986; 28:72–73.
13. Pappas GP. Respiratory protective devices: Rates of medical clearance and causes for work restrictions. *Am J Ind Med* 1999; 35:390–394.
14. Muhm JM. Medical surveillance for respirator users. *J Occup Environ Med* 1999; 41:989–994.
15. Questions and Answers on the Respiratory Protection Standard. Available at *http://www.OSHA.gov/qna.pdf.*
16. U.S. Preventive Services Task Force. Counseling to prevent tobacco use. In *Guidelines from Guide to Clinical Preventive Services,* 2d ed. Baltimore: Williams & Wilkins, 1996.
17. Hauser WA, Hesdorffer DC. *Epilepsy: Frequency, Causes and Consequences.* New York: Demos, 1990.
18. Epilepsy Foundation of America. Driver Licensing. Available at *http://www.efa.org/advocacy/dl_introduction.html.*
19. Hegmann KT. Neurologic disorders. In Hartenbaum NP (ed.), *The DOT Medical Examination: A Guide to Commercial Drivers' Medical Certification.* Beverly Farms, Mass.: OEM Press 1997:104–117.
20. Available at *http://www.efa.org.*
21. Available at *http://www.efa.org/education/handbook.html.*
22. Cryer PE, Polonsky PS. Glucose homeostasis and hypoglycemia. In Wilson JD, Foster DW, Kronenberg HM, Larsen PR (eds.), *Williams' Textbook of Endocrinology,* 9th ed. Philadelphia: Saunders, 1998:939–971.

23. The Diabetes Control and Complications Trial Research Group. The effect of intensive treatment of diabetes on long-term complications in insulin-dependent diabetes mellitus. *New Engl J Med* 1993; 329:977–986.
24. U.S. Department of Transportation, Federal Highway Administration. *Conference on Diabetic Disorders and Commercial Drivers*. Publication No. FHWA-MC-88-041. Washington: Office of Motor Carriers, 1988.
25. Hegmann KT. Endocrine disorders. In Hartenbaum NP (ed.), *The DOT Medical Examination: A Guide to Commercial Drivers' Medical Certification*. Beverly Farms, Mass.: OEM Press, 1997:118–129.
26. Cox D, Gonder-Frederick L, Clarke W. Driving decrements in type I diabetes during moderate hypoglycemia. *Diabetes* 1993; 42:239–243.
27. Clarke WL, Cox DJ, Gonder-Frederick, LA, Kovatchev B. Hypoglycemia and the decision to drive a motor vehicle by persons with diabetes. *JAMA* 1999; 282:750–754.
28. Vaile HG, Pyke DA. Diabetes mellitus and thyroid disorders. In Cox RAF, Edwards FC, McCallum RI (eds.), *Fitness for Work: The Medical Aspects*, 2d ed. Oxford, England: Oxford Medical Publications, 1995:219–230.
29. National Institutes of Health Consensus Development Conference. Health implications of obesity. *Ann Intern Med* 1985; 103:977.
30. Kuczmarski RJ, Flegal KM, Campbell SM, et al. Increasing prevalence of overweight among U.S. adults: The national health and nutrition examinationination surveys. *JAMA* 1994; 272:205–211.
31. Pi-Sunyer FX. Obesity. In Bennett JC, Plum F (eds.), *Cecil Textbook of Medicine*, 20th ed. Philadelphia: Saunders, 1996:1161–1168.
32. Khosla R, Guntupalli KK. Heat-related illnesses. *Crit Care Clin* 1999; 15:251–263.
33. Delaney KA, Goldfrank LR. Hot and cold work environments. In Rom WN (ed.), *Environmental and Occupational Medicine*, 3d ed. Philadelphia: Lippincott-Raven, 1998:1389–1401.
34. Gardner JW. Risk factors predicting exertional heat illness in male Marine Corps recruits. *Med Sci Sports Exerc* 1996; 28:939–944.
35. NIOSH Criteria Documents. *Criteria for a Recommended Standard Occupational Exposure to Hot Environments (Revised Criteria 1986)*. DHHS (NIOSH) Publication No. 86-113. Washington: National Institute of Occupational Safety and Health, 1986.
36. Jackson AS. Prediction of functional aerobic capacity without exercise testing. *Med Sci Sports Exerc* 1990; 22:863–870.
37. Available at *http://janweb.icdi.wvu.edu/english/pubs/OtherPubs/Diabetes.html*.

38. Gergen PJ, Turkeltaub PC, Kaovar MG. The prevalence of allergic skin reactivity to eight common allergens in the U.S. population: Results from the Second National Health and Nutrition Examinationination Survey. *J Allergy Clin Immunol* 1987; 800:669–679.
39. Centers for Disease Control and Prevention. *Vital and Health Statistics: Current Estimates from the National Health Interview Survey, 1994.* Publication No. FHWA-MC-88-041. Washington: U.S. Department of Health and Human Services, Public Health Service, National Center for Health Statistics, December 1995.
40. Centers for Disease Control and Prevention. Surveillance for asthma—United States, 1960–1995. *MMWR* 1998; 47(SS-1).
41. Bush RK. Laboratory animal allergy. *J Allergy Clin Immunol* 1998; 102:99–112.
42. Taivainen AI, Tukiainen HO, Terho EO, Husman KR. Powered dust respirator helmets in the prevention of occupational asthma among farmers. *Scand J Work Environ Health.* 1998; 24:503–507.
43. Elmer KB, George RM. Contact urticaria to the MCU-2A/P gas mask. *Milit Med* 1999; 164:377–378.
44. Available at *http://www.osha-slc.gov/Preamble/RP_html/RESPIRATORY7.html.*
45. Robbins DB. Psychiatric conditions in worker fitness and risk evaluation. In Himmelstein JS, Pransky GS (eds.), Worker Fitness and Risk Evaluations. *Occup Med* 1988; 3:309–321.
46. Available at *http://janweb.icdi.wvu.edu/english/pubs/OtherPubs/PSYCHIATRIC.html.*
47. Duncan HJ, Smith DV. Clinical disorders of olfaction: A review. In Doty RL (ed.), *Handbook of Olfaction and Gustation.* New York: Marcel Dekker, 1995:345–365.
48. Doty R, Monroe C. Occupational screening of olfactory function. In *Occupational and Environmental Chemical Hazards.* Chichester, England: Ellis Horwood Limited, 1987.
49. Hodous T, Petsonk L, Boyles C, et al. Effects of added resistance to breathing during exercise in obstructive lung disease. *Am Rev Respir Dis* 1983; 44:943–948.
50. Muller-Wening D, Neuhauss M. Protective effect of respiratory devices in farmers with occupational asthma. *Eur Respir J* 1998; 12:569–572.
51. Hodous TK. Screening prospective workers for the ability to use respirators. *J Occup Med* 1986; 28:1074–1080.
52. Conrad SA. Pneumothorax and chest wall trauma. In Bone RC, Dantzker DR, George RB, et al. (eds.), *Pulmonary and Critical Care Medicine.* St. Louis: Mosby–Year Book, 1989:O3-1-14.

53. Gobbel WG Jr, Rhea WG Jr, Nelson IA, et al. Spontaneous pneumothorax. *J Thorac Cardiovasc Surg* 1963; 46:331–345.
54. Celli BR. Diseases of the diaphragm, chest wall, pleura, and mediastinum. In Bennett JC, Plum F (eds.), *Cecil Textbook of Medicine,* 20th ed. Philadelphia: Saunders, 1996:448.
55. Available at *http://www.cdc.gov/nchswww/fastats/heart.htm.*
56. Sacco RL. Pathogenesis, classification, and epidemiology of cerebrovascular disease. In Rowland LP (ed.), *Merritt's Textbook of Neurology,* 9th ed. Baltimore: Williams & Wilkins, 1995:227–243.
57. Friedewald WT. Epidemiology of cardiovascular diseases. In Bennett JC, Plum F (eds.), *Cecil Textbook of Medicine,* 20th ed. Philadelphia: Saunders, 1996:171–172.
58. U.S. Department of Transportation, Federal Highway Administration, Office of Motor Carriers. *Conference on Cardiac Disorders and Commercial Drivers*. Publication No. FHWA-MC-88-040. Washington: USDOT, 1987.
59. Gutgesell HP, Atkins DL, Day RW. Common cardiovascular problems in the young: II. Hypertension, hypercholesterolemia and preparticipation screening of athletes. *Am Fam Physician* 1997; 56:1993–1998.
60. Braunwald E. The history. In Braunwald E (ed.), *Heart Disease: A Textbook of Cardiovascular Medicine,* 5th ed. Philadelphia: Saunders, 1997:11–12.
61. Temte JV. Cardiovascular conditions in worker fitness and risk. In Himmelstein JS, Pransky GS (eds.), Worker Fitness and Risk Evaluations. *Occup Med* 1988; 3:241–254.
62. Hartenbaum NP. Cardiovascular disorders. In Hartenbaum NP (ed.), *The DOT Medical Examination*. Beverly Farms, Mass.: OEM Press, 1997:64–81.
63. American College of Sports Medicine. *ACSM's Guidelines for Exercise Testing and Prescription,* 5th ed. Baltimore: Williams & Wilkins, 1995.
64. Zipes DP, Garson A Jr. 26th Bethesda Conference: Recommendations for determining eligibility for competition and athletes with cardiovascular abnormalities. Task Force 6: Arrhythmias. *J Am Coll Cardiol* 1994; 24:892.
65. Myerburg RJ, Castellanon A. Cardiac arrest and sudden cardiac death. In Braunwald E (ed.), *Heart Disease: A Textbook of Cardiovascular Medicine,* 5th ed. Philadelphia: Saunders, 1997:742–756.
66. Treadway KK. Screening for hypertension. In Goroll AH, May LA, Mulley AG (eds.), *Primary Care Medicine,* 3d ed. Philadelphia: Lippincott-Raven, 1995:69–72.
67. Cheitlin MD, Douglas PS, Parmley WW. 26th Bethesda Conference:

Recommendations for determining eligibility for competition in athletes with cardiovascular abnormalities. Task Force 2: Acquired valvular heart disease. *J Am Coll Cardiol* 1994; 24:874–880.

68. Available at *http://janweb.icdi.wvu.edu/english/pubs/OtherPubs/Heart.html.*
69. Gilmore TM, Alexander BH, Mueller BA, et al. Occupational injuries and medication use. *Am J Ind Med* 1996; 30:234–239.
70. Rennie IG, Beeley L. Medication. In Cox RAF, Edwards FC, McCallum RI (eds.), *Fitness forWork: The Medical Aspects,* 2d ed. Oxford, England: Oxford University Press, 1995:45–59.
71. Zelnick SD, McKay RT, Lockey JE. Visual field loss while wearing full face respirator protection. *Am Ind Hyg Assoc J* 1994; 55:315–321.
72. American College of Occupational and Environmental Medicine. The use of contact lenses in an industrial environment: A position statement, 1997. Available at *http://www.acoem.org/paprguid/papers/clens.htm.*
73. U.S. DOL Memorandum to Regional Administrators Regarding Contact Lenses Used with Respirators, 29 CFR 1910.34(5)(ii), February 1988.
74. Available at *http://www.lowvision.org.*
75. Ronk R, White MK. Hydrogen sulfide and probabilities on "inhalation" through a tympanic membrane defect. *J Occup Med* 1985; 27:337–340.
76. Available at *http://janweb.icdi.wvu.edu/english/pubs/OtherPubs/HEARING.html.*
77. King PM. A critical review of functional capacity evaluations. *Phys Ther* 1998; 78:852–866.
78. Available at *http://janweb.icdi.wvu.edu/english/pubs/AccommodationIdeaList.html.*

5

Special Problems with Respirator Use

This chapter reviews some special accommodations integral to respirator use for some individuals. Some physiologic conditions and facial configurations can interfere with the facial seal necessary for a tight-fitting respirator. Occasionally, PLHCPs may prescribe respirators to accommodate workers with heightened susceptibilities to workplace airborne hazards.

Problems with Facial Seal

The Occupational Safety and Health Administration (OSHA) requires that employers conduct fit testing for employees prior to the initial mandated use of any tight-fitting respirator. Employers must repeat the fit testing whenever the employee, employer, physician, licensed health care professional (LHCP), supervisor, or respiratory protection administrator observes changes in the employee's physical condition that could affect respirator fit. 29 CFR 1910.134(g)(1)(i) states that the employer shall not permit the use of tight-fitting facepieces by any employee who has a condition that interferes with valve function or the face-to-facepiece seal. This requirement applies to both negative- and positive-pressure respiratory protective devices that rely on the principle of forming a tight facial seal.

Dental conditions and facial scarring, deformity, or some normal configurations may preclude a correct fit. In such cases, employers or employees may ask the PLHCP for advice. Workers may elect to fix some cosmetic or dental problems. When a medical condition causes a fit problem that cannot or will not be corrected, the PLHCP may recommend a loose-fitting respirator, which the employer must then provide.

Based on numerous studies, OSHA has concluded that facial hair can significantly interfere with the effectiveness of a tight-fitting respirator.[1] The agency therefore prohibits an employer from allowing employees to use tight-fitting respirators if facial hair or jewelry comes between the sealing surface of the facepiece and the face or if it interferes with valve function. Although good industrial hygiene practice dictates a clean-shaven face, OSHA does not prohibit voluntary use of a tight-fitting respirator over facial hair. The employer may mandate that its employees choose jewelry and grooming styles to allow for proper respirator seal.

About 50 to 75 percent of African-American men and 3 to 5 percent of Caucasian men who shave develop pseudofolliculitis barbae of varying severity.[2] In such men, loose-fitting respirators usually will alleviate exacerbations of this problem, but the employer may dictate its own policies regarding the provision of respirators for this purpose.[3] Most men also can use medical and personal hygiene strategies to allow shaving with minimal dermatologic reaction.

Medical Prescription of Respirators for Susceptible Individuals

The PLHCP may be asked to evaluate an employee who has voluntarily chosen to wear a respirator. The OSHA respiratory protection standard defines *voluntary use* as the elective use of a respirator by an employee where such use is not required by OSHA or the employer. Before permitting an employee to use a respirator under these circumstances, OSHA mandates that the employer determine that this use in itself will not create a hazard. In addition, the standard states that the "employer must establish and implement those elements of a written respiratory protection program necessary to ensure that any employee using a respirator voluntarily is medically able to use that respirator and that the respirator is cleaned, stored, and maintained so that its use does not present a health hazard." OSHA exempts the voluntary use of filtering facepieces (dust masks) from this requirement.

The basic approach to assessing an individual who has elected to use a respirator parallels that described in the preceding chapter with one critical difference. If the individual is using the respirator primarily for comfort and peace of mind to further reduce exposures, the consequences of respirator noncompliance and intolerance are of little medical consequence. Accordingly, a PLHCP can recommend approval using more relaxed medical standards.

Some people are susceptible to the adverse effects of an agent below concentrations that do not trouble most people. The medical prescription of a respirator to protect a susceptible individual deserves thoughtful consideration and, in some cases, *more* stringent guidelines for respirator approval than those used for the population at large. For this reason, we recommend that advanced practitioners be involved in assessing individuals whose personal clinician has prescribed respirator use. Certain susceptible workers may benefit from the added protection of a respirator. Individuals with respiratory conditions such as asthma or rhinosinusitis that make them more susceptible to airborne irritants may benefit from using a respirator to minimize their exposures. A simple cellulose-fabric facemask may prevent cold-induced respiratory symptoms.[4] In some circumstances, workers with diseases that involve target organs of occupational toxicants appropriately may use a respirator for added protection. Such conditions may include pulmonary, liver, renal, or neurologic diseases. Respirators also have been used successfully as part of a comprehensive program to reduce the risk that atopic individuals will become sensitized to occupational allergens.[5]

Generally, a respirator should be the last choice to protect a susceptible individual. Employers should implement the usual industrial hygiene hierarchy to minimize a susceptible individual's exposures. The first choices for protection include product substitution, engineering interventions, and employee transfer to areas distant from the hazard. On the other hand, since respirator protective devices reduce an employee's airborne exposures, health care providers appropriately may recommend respirator use as an accommodation for some susceptible individuals who might not otherwise tolerate a work environment. The consequences of respirator failure, noncompliance, or intolerance for such susceptible individuals can become much more serious than for healthy workers—serious enough that in some cases respirator use for this purpose is simply *not* appropriate.

Respirators generally should not be recommended for a worker to continue to work in an environment with potential exposures to substances that have caused his or her illness. OSHA will not allow respirator use for protection of a lead-poisoned worker with ongoing high lead levels. Voluntary respirator use assumes that exposures are already below OSHA exposure limits. The goal of voluntary respirator use is to further reduce the risk associated with ambient toxicant concentrations below the permissible exposure limits. Respirators will not eliminate opportunities for exposure, however. Workers with occupational diseases have already demonstrated a susceptibility to a toxic substance, and the involved target organ is now presumably more compromised. Every effort should be made to eliminate further exposure to the offending toxicant. Hepatologists caring for individuals with chronic liver disease recommend complete abstinence from alcohol, not simple reduction of intake. Exposure to other hepatotoxicants similarly should be eliminated whenever possible.

In some situations, after fully educating an employee about the potential consequences of further exposure, a PLHCP should approve respirator use for added protection in the environment that made the worker ill. If the illness was transient, without sequelae, there is no reason to withhold approval. Exceptions for people with a chronic occupational disease may be appropriate when the illness has been caused by exposure to high concentrations of an agent with a clear dose-response relationship. The voluntary use of an appropriate respirator presumably should more than adequately protect an individual with a mild pneumoconiosis that does not interfere with respirator tolerance. On the other hand, incidental, minimal exposures to offending antigens may be disastrous for individuals with occupational asthma and allergic conditions.

Although many employers offer leave or transfer to pregnant women to minimize hazardous fetal exposures, the U.S. Supreme Court has ruled that pregnant women have the right to make their own decisions regarding fetal protection from occupational hazards.[6] Therefore, a pregnant woman may elect to continue to work in an environment that is potentially hazardous to her fetus. She may desire to use a respirator for a heightened level of protection.

We recommend close monitoring of employees who use respirators to reduce their risk of aggravating an existing condition. The PLHCP should suggest the proper heightened surveillance program. In most cases, this program will be implemented by the attending physician, but it should be coordinated with an occupational health practitioner knowledgeable about the individual's workplace. The monitoring may involve not only indicators of target organ function, such as liver enzymes or pulmonary function, but also, when possible, biologic measures of the toxicant of concern to assess the effectiveness of the respiratory protection.

Interview

The PLHCP reviewing a medical recommendation for the respiratory protection of a susceptible individual should procure sufficient information from the attending practitioner to clearly understand the nature and severity of the worker's medical condition (see Figure 4-1). Data to review include

- Date of symptom onset, cause, and diagnosis
- Results and date of diagnostic tests
- Episodic or chronic symptoms
- Frequency of episodes over the past 2 years, if episodic, with date of last episode
- Current symptoms
- Factors that trigger exacerbations
- Medications and other treatments, including hospitalizations
- Functional impairment, including existing occupational restrictions, history of lost work due to pulmonary illness, impact on exercise or other strenuous activities, impact on activities of daily living
- History of respirator use and tolerance
- Hazards the respirator is meant to minimize

To these data the PLHCP should add knowledge of the work environment in which the individual proposes to use the respirator.

Assessment

The PLHCP should work closely with the company, the attending practitioner and the worker to clearly assess the appropriateness of voluntary respirator use for an individual. With reference to the general guidance outlined earlier and in Chapter 4, the PLHCP should decide whether the person can use a respirator comfortably and safely. To enhance the likelihood of successful respirator use, the PLHCP should consider the underlying medical problem, along with the occupational exposures, when recommending a specific device. For example, a powered air-purifying respirator (PAPR) will offer better protection and comfort for an individual with a chronic cough than a tight-fitting negative-pressure respirator.

An affirmative answer to any of the following questions may indicate the need for respirator denial or restrictions:

- Are the worker's symptoms likely to provoke a break in the facial seal of a tight-fitting respirator?
- Are the symptoms likely to provoke noncompliance with respirator use? Is the worker likely to remove the respirator to

dispose of sputum, to administer medications, or to decrease respiratory load?

- Are the symptoms sufficient to place substantial physiologic limitations on the person's ability to perform activities similar to those which will be required while wearing a respirator?
- Does the person have a history of respirator intolerance due to the underlying medical problem?
- Will noncompliance with respirator use be likely to precipitate significant medical consequences?

Plan

Respirator Approval Decision for Voluntary Use

1. Individuals who choose to use a filtering facepiece (dust mask): Company may allow use without medical approval.
2. Individual who chooses to wear any other type of respirator for comfort and peace of mind:
    - Assess using guidelines in Chapter 4.
    - Approve for respirator use without additional consultation, unless a medical condition is identified that requires further evaluation.
    - Recommend reevaluation by brief questionnaire (see Table 3-3) or age-related guideline (see Table 3-4), barring any change in health status or working conditions.
3. Individual who requests respirator use as an accommodation for an underlying medical condition:
    - Ascertain whether more effective industrial hygiene measures are feasible to reduce exposures of concern.
    - Obtain written recommendations for respirator use from the worker's attending physician.
    - Ensure correct respirator selection with accommodations for any special needs related to the medical problem.
    - Use guidelines appropriate to the medical problem to assess the worker's ability to use a respirator while performing assigned job tasks.

- Consider the hazards of the work environment, the consequences of respirator or job performance failure, the history of respirator tolerance, and the functional impact of the worker's disease.
- If respirator use is approved, recommend heightened medical surveillance of the underlying condition and, when possible, biologic monitoring of the agent(s) of concern to assess the effectiveness of the respirator.
- Recommend reevaluation based on the severity and progression of the disease as well as confidence in the initial recommendations.

Education

1. Review the risks of accidental or incidental exposures.
2. Report for reevaluation for respirator use if difficulty tolerating respirator or performing job.
3. Report exacerbations of underlying medical problem.
4. Counsel about the importance of avoiding agents that aggravate the condition at home, as well as at work.

Conclusion

The employer or employee may ask the PLHCP for advice about respirator selection and use because of medical problems or facial shapes that pose difficulties with respirator fit. In some cases, medical treatment may allow a good fit. In other cases, alternative respirators may be necessary to achieve adequate respiratory protection. Employers may allow voluntary use of dust masks without medical review. The voluntary use of any other type of respirator requires medical evaluation. When respirators are used primarily for comfort, PLHCPs may approve elective use, barring any identified medical problems. The prescription of a respirator as an accommodation for a susceptible individual, however, demands the involvement of an advanced practitioner of occupational medicine to consider the goals and risks of the use of personal protective equipment for this purpose.

References

1. Available at *http://www.osha-slc.gov/Preamble/RP_html/RESPIRATORY7.html.*
2. Habif TB. *Clinical Dermatology: A Color Guide to Diagnosis and Therapy,* 3d ed. St. Louis: Mosby–Year Book, 1996:248–249.
3. Available at *http://www.osha-slc.gov/OshDoc/Interp_data/I19961003.html.*
4. Millquist E. Prevention of asthma induced by cold air by cellulose-fabric face mask. *Allergy* 1995; 50:221–224.
5. Fisher R. Prevention of laboratory animal allergy. *J Occup Environ Med* 1998; 40:609–613.
6. *International Union et al.* v. *Johnson Controls, Inc.,* 59 USLW 4209, March 20,1991.

6

Communicating Respirator Use Recommendations

The Occupational Safety and Health Administration (OSHA) requires the employer, not the physician or licensed health care professional (PLHCP), to finally determine an employee's ability to use a respirator. To make this decision, most employers rely in large part on the PLHCP's written recommendations. This chapter explains how to transmit these recommendations. With reference to some common case scenarios, we offer some specific advice about language for suggesting restrictions or accommodations.

In the final step of the medical evaluation process, the PLHCP provides the employer and employee with a written opinion regarding the employee's ability to use a respirator. Specifically, 29 CFR 1910.134(e)(6) requires that the opinion transmit only the following information:

1. Whether the employee may use a respirator and, if so, whether there are any limitations relating to either the medical condition of the employee or the workplace conditions in which the respirator will be used
2. The need, if any, for follow-up medical evaluations
3. A statement that the PLHCP has provided the employee with a copy of the PLHCP's recommendations

Medical confidentiality statutes prohibit the PLHCP from providing the employer with diagnoses or other medical details of an employee's condition without permission. The employer may not maintain the completed questionnaires or PLHCP's findings unless the employer's health and central administrative offices are separate.

After either the questionnaire or the medical examination approach, the PLHCP may elect to (1) medically qualify the employee for all respirators, (2) medically qualify the employee for selected respirators, (3) medically disqualify the employee for all or selected respirators, (4) identify job task or environmental limitations, (5) recommend further evaluation prior to concluding the medical determination, or (6) suggest periodic reevaluations. The Respirator Use Recommendation Form (Figure 6-1) allows easy recording of these medical opinions and the type of evaluation that forms the basis of the recommendations. The form also provides the opportunity to document the mandated transmittal of the respirator use recommendations to the employee.

The PLHCP should bring clinically significant findings to the attention of the employee with recommendations for appropriate follow-up. If these recommendations cannot be reviewed in person, the PLHCP must either mail them directly to workers' home or ask the employer to distribute the sealed letters to employees at the work site.

The clinician may qualify an individual for one respirator but not for others. Many PLHCPs require more extensive medical evaluation before qualifying a worker for self-contained breathing apparatus (SCBA) use than for other respirators. In such cases, rather than approving an individual for all respirators, the PLHCP may want to specifically withhold approval for SCBA use despite the absence of any clear disqualifying information. Figure 6-1 provides a simple check box to transmit this information. Second, the PLHCP may determine that an individual with a specific medical condition should not use respirators with high inspiratory resistance, such as a negative-pressure air-purifying respirator. In this case, the PLHCP communicates that the worker may wear a PAPR or a positive-pressure atmosphere-supplying respirator but may not wear a negative-pressure respirator. Figure 6-1 provides an easy checklist for selective respirator approval.

In addition to selectively approving workers for specific respirators, the PLHCP may suggest limitations on the employee's use of a respirator. These restrictions must be communicated in writing to both the employer and the employee. The following four cases illustrate an approach to generating recommendations for common clinical conditions encountered during the medical evaluation for respirator use. Each case describes the worker's job, medical evaluation, assessment, and plan. A completed Respirator Use Recommendation Form accompanies the case to depict the information transmitted to the employer. The medical condition noted in the figure heading would not be included in any communication with the employer. Just as occupational health professionals routinely consider and, when appropriate, document the need for job restrictions at the conclusion of

Figure 6-1. Respirator Medical Recommendation

Date: ____________ Company: ____________________ Job Title: ____________

Name: ______________________________ Social Security: ________________

Based on review of the OSHA Respirator Health Questionnaire this individual is:

_____ Medically approved for all respirator(s) with the exception of SCBA, and subject to fit test.

Based on interview, physical examination and further evaluation as appropriate, this individual is:

_____ Medically approved for all respirator(s) including SCBA, and subject to fit test

_____ Medically approved for only the following types subject to satisfactory fit test:
- _____ Dust mask
- _____ Negative Pressure
- _____ Powered air purifying respirator
- _____ Supplied air
- _____ Self-contained breathing apparatus (SCBA)

_____ Employee may decline respirator-requiring assignments for temporary health related difficulties.

_____ Respirator assignment must not be for IDLH (Immediate Danger to Life or Health) environments.

_____ Employees should not be expected to perform rescue duty or serve as a member of a rescue team. If able to wear a respirator at the time, then rescue duties may be performed.

_____ Require further medical information/evaluation prior to qualifying for respirator use.

_____ Other recommendations and suggested accommodations:

Recommended time period for next exam:

[ ] annual [ ] 2 years [ ] 5 years [ ] _____

Employee has been provided with a copy of this written recommendation.

Yes [ ] No [ ]

______________________________ ____________________

Physician/Practitioner signature Date

a preplacement examination, they should apply similar thought when communicating with employers about respirator use.

Case 1: J. Doe (Hypertension)

Job title: Maintenance and repair

Environmental Exposures and Respirator Use

The job requires the cleaning of nonhazardous waste from an underground catch well on a biweekly basis. The work area is a confined and oxygen-deficient space (immediately dangerous to life or health, IDLH). Required work expenditure is estimated at 8 to 9 metabolic equivalents (METs) over 2 to 4 hours with breaks. Required personal protective equipment (PPE) includes supplied-air full-facepiece respirator with an auxiliary self-contained unit, water-resistant outer garment, calf-height water-resistant boots, three-quarter-length rubber gloves, and hearing protection. Environmental temperatures range between 70 and 90°F depending on the season.

Respirator Evaluation

Due to the level of work required and the IDLH environment, the employer chose the examination option for respirator medical evaluation. The company contracts with the local occupational health clinic for the evaluations.

Health History

Mr. Doe is a 48-year-old man who reports a history of hypertension. He was treated initially with medication, but treatment was discontinued approximately 5 years ago after he lost 25 lb. Since that time, he has regained 20 lb. The last visit to his primary care provider was over 4 years ago. He denies any cardiovascular symptoms but does complain of an early morning productive cough, which he relates to cigarette smoking (1/2-pack per day × 20 years). Sputum is white to light yellow in color. He denies additional respiratory symptoms. He has five cardiac risk factors (i.e., male, age, cigarette smoking, history of hypertension, and obesity). Family history is unremarkable for cardiac disease. Review of systems is negative. Past medical history is unremarkable, with the exception of his hypertension. Mr. Doe reports a sedentary lifestyle and is engaged in no regular exercise program. In other jobs, he has worn both air-purifying and supplied-air respirators for short durations without a problem, e.g., 45 to 60 minutes. Last respirator use was approximately 3 years ago. His job tasks previously have not included cleaning the catch well, but because of a staffing shortage, job tasks were reassigned.

Medical Examination

The medical examination focuses on those systems identified in the history which may have an impact on the worker's ability to wear the selected respirator in an IDLH environment.

> *Examination:* BP, 158/96 mmHg; repeat BP at 20 minutes, 162/98 mmHg; HR, 84; R 12; height, 72 inches; weight, 230 lb
>
> *Eyes:* Funduscopic examination reveals arteriolar narrowing.
>
> *Heart:* Regular rhythm; normal S_1 and S_2, without murmurs, gallops, or rubs; no jugular venous distension
>
> *Lungs:* Clear bilaterally

Additional Testing

> *Spirometry:* Mild obstructive lung disease
>
> *Baseline electrocardiogram (ECG):* Normal
>
> *Urine:* All indices are negative, including protein

History and examination reveal stage I hypertension with no indication of target-organ damage, except for mild hypertensive retinopathy. Spirometry reveals mild airways obstruction probably consistent with mild chronic obstructive pulmonary disease due to his lifelong tobacco habit. None of these conditions contraindicates respirator use. However, concern about his multiple cardiac risk factors and the additional cardiovascular demands associated with the job's physical requirements, respirator use, and thermal conditions prompts the physician to order an exercise stress test (EST).

EST. No abnormal ECG or clinical changes are observed. Test is stopped due to patient fatigue at 7 METs. Test is conducted without the patient wearing PPE.

The EST result has diminished a concern for any imminent health threat. The employee's physical deconditioning raises questions as to whether

Mr. Doe can meet the job requirements, particularly when working in full personal protective gear. A simulated work trial is ordered.

Work Trial. The simulated test results indicate worker fatigue at 1/2-hour with perceived exertion of 8 on the Rating of Perceived Exertion (RPE) scale. Vital signs remained stable, although mild hypertension persists.

Assessment

The tiered evaluation identifies no immediate health threat. Mr. Doe, however, is hypertensive and deconditioned and does not have the physical stamina to meet the job requirements. He remains at risk for cardiac disease and further respiratory compromise. He does not have appropriate health care.

Plan

1. Medically qualify for all respirators but unable to clean the catch well without the following accommodation: a gradual increase in catch-well work hours over a 6- to 8-week period; e.g., begin with less strenuous activities and increase by one half-hour time increments (Figure 6-2).
2. Follow up with personal physician (all records forwarded with letter reviewing work requirements and recommendations).
3. Recommend exercise prescription by primary clinician.
4. Encourage smoking cessation.
5. Encourage weight loss.
6. Review cardiovascular burden of respirator use with employee.
7. Instruct employee to report signs and symptoms that should trigger medical reevaluation.
8. Reevaluate in 1 year.
9. Instruct to bring copies of interval primary care records to next year's evaluation.
10. Provide employee with a written copy of the respirator use recommendation.

Figure 6-2. Case 1: Hypertension

Date: 10/2/99 Company: ABC Company Job Title: Maintenance

Name: J. Doe Social Security: 000-00-0000

Based on review of the OSHA Respirator Health Questionnaire this individual is:

_____ Medically approved for all respirator(s) with the exception of SCBA, and subject to fit test.

Based on interview, physical examination and further evaluation as appropriate, this individual is:

__x__ Medically approved for all respirator(s) including SCBA, and subject to fit test

_____ Medically approved for only the following types subject to satisfactory fit test:
- _____ Dust mask
- _____ Negative pressure
- _____ Powered air purifying respirator
- _____ Supplied air
- _____ Self-contained breathing apparatus (SCBA)

_____ Employee may decline respirator-requiring assignments for temporary health related difficulties.

_____ Respirator assignment must not be for IDLH (Immediate Danger to Life or Health) environments.

_____ Employees should not be expected to perform rescue duty or serve as a member of a rescue team. If able to wear a respirator at the time, then rescue duties may be performed.

_____ Require further medical information/evaluation prior to qualifying for respirator use.

__x__ Other recommendations and suggested accommodations:

Will require a period of acclimatization to clean "the catch well"—recommend 1/2 *hour work period with respirator followed by rest period of low intensity work for* 1 *hour. Grdually increase work time in* 1/2 *hour intervals each week. If unable to tolerate refer for further medical evaluation.*

Recommended time period for next exam:

[x] annual [ ] 2 years [ ] 5 years [ ] _____

Employee has been provided with a copy of this written recommendation.

Yes [x] No [ ]

________________________ ________________

Physician/Practitioner signature Date

Case 2: M. Brown (Asthma)

Job title: Spray painting

Environmental Exposures and Respirator Use

The job requires spray painting of various products in a paint booth. Paints used are water-based and contain no isocyanates or sensitizers. Required work expenditure is between 4 and 5 METs. Required personal protection includes a negative-pressure, air-purifying half-facemask respirator, safety glasses, Tyvek outer garment, gloves, and water-resistant boots. Ambient temperature ranges from 70 to 90°F, depending on the season.

Respirator Evaluation

The company uses the OSHA questionnaire for the medical evaluation for respirator use. Within her scope of practice, the company's occupational health nurse (LHCP), administers, reviews, and recommends respirator use. In this case, affirmative responses on the questionnaire prompt the LHCP follow-up interview and further evaluation.

Health History

Ms Brown is a 35-year-old woman who has applied for a transfer to the paint line. On the OSHA questionnaire, she reports a history of asthma. The following history is obtained on further questioning by the LHCP. Ms Brown was diagnosed with asthma 10 years ago. Asthma triggers include cat dander, certain springtime pollens, and strenuous exercise. She reports that her asthma is under good control with her current, daily pharmaceutical treatment, though she notes an occasional need for additional inhaler use when exposed to antigens. During an attack, symptoms may include chest tightness, cough, and wheezing. Her last attack was 1 month ago. She reports it as mild and easily controlled with her medication. She has never been hospitalized, nor has she had any uncontrolled episodes that required either an emergency department or physician office visit. She has yearly visits with her primary physician, most recently 3 months ago. She reports that spirometry conducted at that time was "ok." She has no previous history of respirator use. Ms Brown reports that she has used both latex- and oil-based paints in her home without adverse effects. She walks 4 miles

(12-minute miles) three times per week. Past medical history is unremarkable with the exception of asthma.

Current medication: Inhaled beclomethasone diproprionate daily and albuterol inhaler as needed

Medical Examination

The company's occupational health clinic procedure for further respirator evaluation requires baseline vital signs and a physical assessment appropriate to the health concern.

Examination: BP, 110/72 mmHg; HR, 66; R, 12; height, 66 inches; weight, 120 lb

Skin: Color good

Heart: Regular rhythm; normal S_1 and S_2; without murmurs, gallops, or rubs

Lungs: Clear and without wheezes on forced expiration

Medical Record Review

Spirometry report from primary care provider notes mild obstruction reversible with bronchodilation. The medical records note good control of asthma with current treatment regime. The primary care provider indicates that there are no contraindications to respirator use.

Assessment

Ms Brown's asthma is stable and under good control with the current medical regime. Paint odors historically have not triggered her asthma. Documentation from her primary care provider indicates no medical contraindication to respirator use. A review of work exposures indicates no known sensitizers. Ms Brown has a good understanding of her asthmatic condition, early warning symptoms, and treatment regime. She should be able to wear the selected respirator and perform her job tasks without problem. She may not tolerate respirator use, however, while in the midst of an asthmatic attack.

Plan (per LHCP, Depending on the Scope of Practice, or an Appropriate Advanced Clinician)

1. Medically qualify worker for use of the selected respirator but recommend that she have the option of declining use for temporary health-related problem. Would not qualify her for air-supply or self-contained breathing apparatus (SCBA) without medical review of new work tasks (Figure 6-3).
2. Instruct worker to report any respiratory problems to the health office.
3. Allow breaks for medication use as needed.
4. Identify a clean place to keep asthma medication.
5. Follow up with the occupational health nurse in 6 weeks for reevaluation (brief interview). If no problems at that time, recommend annual follow-up by brief questionnaire (see Table 3-3).
6. Provide worker with a written copy of the respirator use recommendation.

Case 3: M. Green (Diabetes)

Job title: Sandblaster and emergency maintenance response team

Environmental Exposures and Respirator Use

Sandblasting requires the use of a supplied-air respirator with a hood due to silica exposure. Energy expenditure is between 4 and 6 METs. Additional PPE includes safety glasses, coveralls, and hearing protection. Ambient temperature remains constant between 70 and 75°F.

Emergency response requires the use of a SCBA or supplied-air respirator with an auxiliary self-contained unit. Depending on the situation, the job may require work in an IDLH environment under difficult conditions. Anticipated work expenditure is between 6 and 8 METs or greater. Additional PPE required includes chemical-resistant gloves and outer garment. Ambient temperature may range from 70 to 100°F.

Figure 6-3. Case 2: Asthma

Date: 10/2/99 Company: ABC Company Job Title: Spray Painter

Name: M. Brown Social Security: 000-00-0000

Based on review of the OSHA Respirator Health Questionnaire this individual is:

_____ Medically approved for all respirator(s) with the exception of SCBA, and subject to fit test.

Based on interview, physical examination and further evaluation as appropriate, this individual is:

_____ Medically approved for all respirator(s) including SCBA, and subject to fit test

__x__ Medically approved for only the following types subject to satisfactory fit test:

- __x__ Dust mask
- __x__ Negative pressure
- __x__ Powered air purifying respirator
- _____ Supplied air
- _____ Self-contained breathing apparatus (SCBA)

__x__ Employee may decline respirator-requiring assignments for temporary health related difficulties.

__x__ Respirator assignment must not be for IDLH (Immediate Danger to Life or Health) environments.

__x__ Employees should not be expected to perform rescue duty or serve as a member of a rescue team. If able to wear a respirator at the time, then rescue duties may be performed.

_____ Require further medical information/evaluation prior to qualifying for respirator use.

__x__ Other recommendations and suggested accommodations:
Allow breaks for medication use

Recommended time period for next exam:

[ ] annual [ ] 2 years [ ] 5 years [x] *Brief follow-up in 6 weeks*

Employee has been provided with a copy of this written recommendation.

Yes [x] No [ ]

_________________________ _________________

Physician/Practitioner signature Date

Respirator Evaluation

The company uses the OSHA questionnaire as the initial screening tool for medical evaluation for respirator use. The company contracts with a local occupational clinic. Completed questionnaires are forwarded to the clinic for review. At the clinic, an occupational health nurse (OHN) conducts an initial review and approves employees who designate all negative responses. Within her scope of practice, the nurse follows up positive responses by interview. When necessary, she refers individuals to the appropriate advanced clinician.

Health History

Mr. Green is a 28-year-old man who reports diabetes on the questionnaire. A telephone interview with Mr. Green by the nurse reveals that he has a 12-year history of type I diabetes. He initially had problems with insulin regulation, but for the past 5 years he has been under good control. Six months ago he began using an insulin pump. Mr. Green adjusts his dosage daily based on blood sugar readings, diet, and activity. He reports 4 to 5 episodes of mild hypoglycemia per month (lowest blood sugar, 60 mg/dl) and denies any episode for which he has required assistance for over a year. He demonstrates good knowledge of his disease and treatment.

Mr. Green runs 3 miles per day (8-minute miles). He has worn an air-purifying respirator in the past but never an SCBA. He has volunteered for the emergency response team.

At the end of the interview, the occupational health nurse (OHN) discusses with Mr. Green the need for further information from his primary care provider. With permission from Mr. Green, the release of information form is faxed to his attention at the company. Mr. Green returns the signed and witnessed form to the clinic.

Medical Examination

None. Documentation from the primary care provider identifies that Mr. Green's past year's fasting blood sugars have ranged from 110 to 160 mg/dl. His most recent HgA1c is 7.5 percent. There is no record of severe hypoglycemia in the last year or evidence of end-organ complications. The provider concurs that the patient has a good understanding of his disease and treatment, although he notes that the patient is a relatively new user of the insulin pump.

Assessment

The LHCP refers the medical records for review by a more advanced practitioner. Mr. Green's history of diabetes warrants separate assessments for each respirator requirement. As a pipe fitter/welder, Mr. Green has successfully worn an air-purifying respirator in the past without problems. A brief modest exposure to silica as the result of respirator failure probably would not cause any serious health effects. The work requires a moderate degree of energy, which he tolerates without problem.

The emergency response requirements pose a concern. The job task's unpredictable energy and time requirements combined with possible thermal load, and the "stress" response places Mr. Green at risk for a hypoglycemic reaction in an IDLH environment. Although he may be able to perform many emergency tasks safely, his medical condition makes his participation unreliable in an emergency.

Plan

1. Medically qualify him for use of the air-supply respirator and SCBA but recommend that he have the option of declining use for a temporary health-related problem. Do not qualify him for SCBA use in an emergency response (Figure 6-4). Avoid respirator use for protection from an IDLH atmosphere.
2. Recommend buddy system during respirator use.
3. Review impact of various work activities on insulin needs and blood sugar.
4. Consider release of information regarding diabetes to company first aid team.
5. Recommend follow-up evaluation at a minimum of 1 year or earlier if a change in treatment occurs.
6. Instruct worker to report signs and symptoms that should trigger medical reevaluation.
7. Instruct worker to bring copies of interval primary care records to next year's evaluation.
8. Provide worker with a written copy of the respirator use recommendation.

Figure 6-4. Case 3: Diabetes

Date: 10/2/99 Company: ABC Company

Job Title: Sandblaster/Response Team

Name: M. Green Social Security: 000-00-0000

Based on review of the OSHA Respirator Health Questionnaire this individual is:

____ Medically approved for all respirator(s) with the exception of SCBA, and subject to fit test.

Based on interview, physical examination and further evaluation as appropriate, this individual is:

__x__ Medically approved for all respirator(s) including SCBA, and subject to fit test

____ Medically approved for only the following types subject to satisfactory fit test:
- ____ Dust mask
- ____ Negative pressure
- ____ Powered air purifying respirator
- ____ Supplied air
- ____ Self-contained breathing apparatus (SCBA)

__x__ Employee may decline respirator-requiring assignments for temporary health related difficulties.

__x__ Respirator assignment must not be for IDLH (Immediate Danger to Life or Health) environments.

__x__ Employees should not be expected to perform rescue duty or serve as a member of a rescue team. If able to wear a respirator at the time, then rescue duties may be performed.

____ Require further medical information/evaluation prior to qualifying for respirator use.

____ Other recommendations and suggested accommodations:

Recommended time period for next exam:

[x] annual [ ] 2 years [ ] 5 years [ ] ____________

Employee has been provided with a copy of this written recommendation.

Yes [x] No [ ]

______________________ ______________

Physician/Practitioner signature Date

Case 4: T. Jones (Pseudofolliculitis Barbae)

Job title: Machinist

Environmental Exposures and Respirator Use

Metal grinding requires the use of a negative-pressure half-mask air-purifying respirator. Exposures include metal dust and metal working fluids. Energy expenditure is between 5 and 7 METs. Ambient temperature usually hovers in the 60 to 70°F range but in the summer can reach the low 90s. Additional PPE includes safety glasses, hearing protection, protective apron, and gloves.

Respirator Evaluation

The company uses the OSHA questionnaire as the initial screening tool for medical evaluation for respirator use. The company contracts with a local occupational health program for onsite occupational nursing and nurse-practitioner services. The occupational health nurse conducts an initial review and approves employees who designate all negative responses. The nurse-practitioner, consulting an occupational physician as necessary, evaluates employees requiring further assessment.

Health History

Mr. Jones is a bearded, 26-year-old African-American man who reports a skin rash on the questionnaire. In the interview with the OHN, Mr. Jones notes that he has had recurrent problems with facial rash in his beard area. The rash usually develops only when he shaves regularly. He saw a family doctor once, who told him he had a shaver's rash. He says the doctor told him to avoid shaving. He reports no history of allergies or other skin problems. Review of systems is negative.

In the past, Mr. Jones has worn a dust mask, which further irritated his face. He says that due to the facial irritation, he would often not wear the mask even though he worked in a very dusty environment. He currently wears a beard and has no complaint of facial rash.

Medical Examination

None. The OHN consults with the nurse-practitioner regarding assessment and plan.

Assessment

Mr. Jones reports a history of facial skin rash, aggravated by shaving and, in the past, use of a dust mask respirator. Work as a metal grinder requires wearing a half-facemask negative-pressure air-purifying respirator for up to a maximum of 3 to 4 hours per day. To wear a tight-fitting respirator, Mr. Jones must shave. Shaving to achieve a tight facial seal puts Mr. Jones at risk for a recurrence of his facial rash. Dermatologic consultation may provide additional strategies that Mr. Jones could employ that would allow him to shave without inciting the folliculitis.

Plan

1. Medically qualify the worker for respirator use, but recommend a powered air-purifying respirator (PAPR) with a hood pending further medical evaluation (Figure 6-5).
2. Obtain medical records to confirm prior skin diagnosis.
3. Recommend follow-up with the primary care provider and possible dermatologist referral if the company's policy discourages provision of a PAPR for this purpose.
4. Instruct worker to report signs and symptoms that should trigger medical reevaluation.
5. Recommend annual reevaluation at time of fit test with short questionnaire (see Table 3-3).
6. Provide worker with a written copy of the respirator use recommendation.

Case 5: Voluntary Use

In some cases, a practitioner may advocate that an individual with a specific medical condition wear a respirator to provide heightened personal protection from occupational exposures not hazardous for most workers. A landscaper with allergic rhinitis may benefit from the use of a disposable filtering respirator. Figure 6-6 illustrates this type of communication with the employer. Another scenario may include a pregnant woman anxious to maintain her job in an area where modest airborne concentrations of solvents exist but who may want to wear a respirator to minimize her exposure. The practitioner may recommend that employers consult the OSHA respiratory protection standard for their responsibilities when they permit voluntary respirator use.

Figure 6-5. Case 4: Pseudofolliculitis Barbae

Date: 10/2/99 Company: ABC Company Job Title: Metal Grinder

Name: T. Jones Social Security: 000-00-0000

Based on review of the OSHA Respirator Health Questionnaire this individual is:

_____ Medically approved for all respirator(s) with the exception of SCBA, and subject to fit test.

Based on interview, physical examination and further evaluation as appropriate, this individual is:

_____ Medically approved for all respirator(s) including SCBA, and subject to fit test

__x__ Medically approved for only the following types subject to satisfactory fit test:

- _____ Dust mask
- _____ Negative pressure
- __x__ Powered air purifying respirator
- __x__ Supplied air (*loose fitting only*)
- _____ Self-contained breathing apparatus (SCBA)

_____ Employee may decline respirator-requiring assignments for temporary health related difficulties.

_____ Respirator assignment must not be for IDLH (Immediate Danger to Life or Health) environments.

_____ Employees should not be expected to perform rescue duty or serve as a member of a rescue team. If able to wear a respirator at the time, then rescue duties may be performed.

_____ Require further medical information/evaluation prior to qualifying for respirator use.

__x__ Other recommendations and suggested accommodations:

Further medical evaluation required to assess employee's capacity to wear a tight-fitting respirator.

Recommended time period for next exam:

[ ] annual [ ] 2 years [ ] 5 years [x] *at annual fit-testing, complete brief questionnaire.*

Employee has been provided with a copy of this written recommendation.

Yes [x] No [ ]

____________________ ____________

Physician/Practitioner signature Date

Figure 6-6. Voluntary Use, Medical Prescription

Date: 10/2/99 Company: ABC Company Job Title: Maintenance

Name: J. Doe Social Security: 000-00-0000

Based on review of the OSHA Respirator Health Questionnaire this individual is:

_____ Medically approved for all respirator(s) with the exception of SCBA, and subject to fit test.

Based on interview, physical examination and further evaluation as appropriate, this individual is:

_____ Medically approved for all respirator(s) including SCBA, and subject to fit test

__x__ Medically approved for only the following types subject to satisfactory fit test:

- __x__ Dust mask
- _____ Negative pressure
- _____ Powered air purifying respirator
- _____ Supplied air
- _____ Self-contained breathing apparatus (SCBA)

_____ Employee may decline respirator-requiring assignments for temporary health related difficulties.

_____ Respirator assignment must not be for IDLH (Immediate Danger to Life or Health) environments.

_____ Employees should not be expected to perform rescue duty or serve as a member of a rescue team. If able to wear a respirator at the time, then rescue duties may be performed.

_____ Require further medical information/evaluation prior to qualifying for respirator use.

__x__ Other recommendations and suggested accommodations.

Employee has a medical condition that may be aggravated by exposure to pollens and dust. Use of a pollen mask will help minimize this problem.

Consult 29 CFR 1910.134 regarding your obligations with respect to the OSHA Respiratory Protection Standard for voluntary use of a dust mask.

Recommended time period for next exam:

[x] annual [ ] 2 years [ ] 5 years [ ] _____

Employee has been provided with a copy of this written recommendation.

Yes [x] No [ ]

_______________________________ _______________

Physician/Practitioner signature Date

Case 6: More Medical Information Needed

At the conclusion of the first step of the medical evaluation, the PLHCP may need additional medical information before completing decision making regarding an individual's respirator use. Rather than withholding communication pending collection of further data, we recommend immediately advising the employer and employee of the need and expected time frame for follow-up medical evaluation. The PLHCP also must transmit recommendations regarding additional evaluations in the future. Refer to Chapter 4 for our recommendations. The Respirator Medical Recommendation form (see Figure 6-1) can be used for these purposes as well.

Summary

Completion of the medical evaluation demands timely, clear communication with the employer and employee of the PLHCP's recommendations about respirator use. This chapter has explained this process and provided sample written determinations for common clinical scenarios.

7

Building Continuous Quality Improvement into the Medical Evaluation for Respirator Use

Following the methods and guidelines of this book will advance the practice of most occupational health providers. Beyond implementing these recommendations, health professionals should build a continuous quality improvement (CQI) process into their respirator program. Ensuring the quality of services demonstrates a commitment to high standards of performance, identifies and measures appropriate outcomes, improves efficiency, and reduces costs. In effect, a continuous quality process demonstrates an ethical commitment to quality care. This chapter briefly reviews the CQI process and applies it to the medical evaluation for respirator use. References in the Further Reading section at the end of this chapter offer more detailed information.

Over the last decade, quality programs in health care have broadened in scope from conventional quality assurance programs. Quality assurance programs focused on conducting tasks and/or activities correctly, with individual staff performance the central point of corrective-action plans. CQI goes beyond individual behavior to address process or systemic issues. CQI identifies the root cause of problems and describes measures to correct the process.[1–3] Implementation of a successful CQI program requires clinic management commitment, staff education, staff and resource commitment, and most notably, staff time.[3] Of the variety of CQI models, we recommend the following approach to continuously improving the medical evaluation of workers for respirator use.

1. *Identify customers (both internal and external).* A *customer* is any individual or group who receives or is affected by an occupational health service. At an employer-based clinic, *internal* customers include employees, supervisors, safety personnel, individuals responsible for fit testing, and human resources staff. The clinic also has "downstream" *external* customers. These customers may place orders to buy goods that

the respirator worker makes. A delay or an inefficiency in the respirator evaluation process can increase the price or timely availability of a product, resulting in a dissatisfied customer. Expanding the definition of an occupational health provider's customers highlights the provider's contribution to a business.[1]

2. *Characterize customer expectations.* This step clarifies specific customer expectations. Most customer groups probably have the same overarching expectation of the program—a timely, rational, inexpensive medical qualification process. Establishing specific parameters for customer expectations can both serve as quality indicators and drive an approach to medical evaluation. If the human resources staff expects medical qualification to occur for a large number of employees within 24 hours, a medical questionnaire approach may best serve the purpose. If, on review, most workers from this particular employer require follow-up that delays the approval process, another approach for collecting clinical data may be indicated.[1]

3. *Outline the process for meeting customer expectations.* This step maps the process needed to meet both professional standards and customer expectations. A flowchart provides one method for standardizing and organizing the service by breaking it into discrete steps. A flowchart also helps identify where clinical or customer expectations cannot be met without adjustments in either the process or in the expectation.[1]

4. *Select quality indicators.* CQI does not aim to measure every aspect of a process. To be effective, CQI focuses on a few critical indicators that reflect the quality of frequent or important events. Appropriate indicators can be chosen from customer expectations, professional standards, and key steps in the process.[1,2] Table 7-1 offers several indicators that occupational health services may wish to use for measuring the quality of their respirator medical evaluations. Timeliness of reporting respirator recommendations to the employer is a typical indicator. By recording the number of respirator assessments that are completed within a defined time period, an auditor can calculate the frequency that the service meets this key customer expectation. The occupational clinic should set a performance threshold that reflects either best or past practice, with the perpetual goal of improvement. Indicators based on OSHA regulations may be assigned the goal of perfect accomplishment. For example, 100 percent of the time a clinic should perform follow-up examinations for employees who have answered OSHA respiratory questions 1 to 8 affirmatively. High-quality service for other indicators appropriately may be associated with lower performance thresholds. Reporting all respirator evaluation results for a large number of employees to the employer within 48 hours will never

Table 7-1. Examples of Quality Indicators and Thresholds

Indicator	Threshold
Documentation/Medical Record	
Availability of employer environmental and respirator usage information prior to medical evaluations	100%[a]
Copy of the employer respiratory protection program	100%[a]
Documentation of employee receipt of respirator recommendation	100%[a]
Clinical Practice	
Documentation of rationale for denial or accommodation for respirator use	90%
Appropriate use of diagnostic testing as defined by clinical guidelines	95%
Customer Service	
Forty-eight-hour turnaround time for respirator medical qualification recommendation	90%
Exceed employer expectations regarding provider communication on delayed cases (survey question)	92%
Exceed employee's expectation regarding information as to the purpose of additional testing	95%

[a]100% threshold is assigned because this is a requirement of the standard.

be attainable because at least some percentage of assessments will require the collection of additional data. Realistically, the clinic management may set a quality threshold for timeliness at a lower number, such as 90 percent. One hundred percent attainment of the 48-hour goal actually may suggest that thorough assessments are not being conducted.

5. *Identify areas for improvement.* Next, the provider organization develops an audit tool based on the chosen quality indicators (Table 7-2). Unless reviewing a small number of respirator assessments, the provider should select a sampling strategy to provide a statistically sufficient, but not overwhelming, amount of data for analysis. A clinic may randomly select 20 percent of charts between, say, January 1, 2000

Table 7-2. Sample Respirator Medical Evaluation Audit Tool

Survey time period (includes all records during the time period): 1/01/2000–4/30/2000
Total number of companies:[a] 50
Random selection percentage of total (minimum of 10 companies[a] surveyed): 20%

Company name:	ABC Co.	Hills Co.	Grant & Co	B&B	Cable	Painters	Monroe	ADD	Lexing	Sun Co.
Indicator										
Availability of current employer environmental and respirator usage information prior to medical evaluation.	✓	✓	✓	✓	✓	✓			✓	✓
Copy of employer respiratory protection program	✓	✓		✓	✓	✓		✓	✓	

Survey Findings

Indicator	Threshold	Survey results
Availability of a current employer environmental and respirator usage information prior to medical evaluations	100%	80%
Copy of employer respiratory protection program	100%	70%

Plan 1: Review process for obtaining both environmental and respirator usage information and company respiratory protection plan.
Plan 2: Identify mechanism to update current information.
[a]May substitute department for company.

and continuing through April 30, 2000. Stakeholders should review and analyze the data. Outcomes falling short of established thresholds represent areas for improvement.[1]

6. *Identify and modify causes of quality problems.* Providers should work with all process holders to identify reasons for failing to meet expected quality thresholds. When factors out of the provider's control affect performance quality, these factors should be brought to the customer's attention for improvement or adjustment of the expectation.[1]

7. *Improve continuously.* When an organization consistently meets the performance thresholds for the chosen indicators, providers may choose to remeasure to validate results, raise the thresholds, or measure other aspects of the process. Measurement, analysis, and improvement continue indefinitely. Since quality may degrade over time, indicators whose thresholds have been met successfully in the past should be reevaluated periodically.[1]

Occupational health professionals will conduct millions of medical evaluations for respirator use. Following a systematic approach to quality improvement minimizes staff and customer frustration with an imperfect process. This book provides an approach for physicians/licensed health care professionals (PLHCPs) to improve the quality of the medical evaluation and opinions generated. In a CQI program, diverse stakeholders help identify all steps in the medical evaluation process, address unintended consequences of improvements, and ease the implementation of changes. CQI furthers the goal of this book by including nonclinical stakeholders in a company's respiratory protection program.

Further Reading

Leebov W, Ersoz C. The Health Care Manager's Guide to Continuous Quality Improvement. Chicago: American Hospital Publishing, 1991. *This comprehensive guide to the practical application of continuous quality improvement offers readers an understanding of CQI concepts, lays out the process-improvement model, and supplies essential tools and techniques for implementing the model.*

Moser R Jr. *Effective Management of Occupational and Environmental Health and Safety Programs: A Practical Guide,* 2d ed. Beverly Farms, Mass.: OEM Press, 1999. *This how-to guide is designed to help occupational health providers plan, implement, and evaluate occupational and environmental health and safety programs.*

References

1. Leebov W, Ersoz C. *The Health Care Manager's Guide to Continuous Quality Improvement*. Chicago: American Hospital Publishing, 1991.
2. Ryder R, Newkirk WL. Integrating continuous quality improvement into occupational health programs. In: Newkirk WL (ed.), *Occupational Health Services Practical Strategies for Improving Quality and Controlling Costs*. Chicago: American Hospital Publishing, 1993:13–21.
3. Salazar MK (ed.). *AAOHN Core Curriculum for Occupational Health Nursing*. Philadelphia: Saunders, 1997:113–116.

Appendix 1

OSHA Respiratory Protection Standard 29CFR 1910.134

This section applies to General Industry (part 1910), Shipyards (part 1915), Marine Terminals (part 1917), Longshoring (part 1918), and Construction (part 1926).

1910.134(a)

(a) *Permissible practice*

(a)(1) In the control of those occupational diseases caused by breathing air contaminated with harmful dusts, fogs, fumes, mists, gases, smokes, sprays, or vapors, the primary objective shall be to prevent atmospheric contamination. This shall be accomplished as far as feasible by accepted engineering control measures (for example, enclosure or confinement of the operation, general and local ventilation, and substitution of less toxic materials). When effective engineering controls are not feasible, or while they are being instituted, appropriate respirators shall be used pursuant to this section.

(a)(2) Respirators shall be provided by the employer when such equipment is necessary to protect the health of the employee. The employer shall provide the respirators which are applicable and suitable for the purpose intended. The employer shall be responsible for the establishment and maintenance of a respiratory protection program which shall include the requirements outlined in paragraph (c) of this section.

1910.134(b)

(b) *Definitions.* The following definitions are important terms used in the respiratory protection standard in this section.

Air-purifying respirator means a respirator with an air-purifying filter, cartridge, or canister that removes specific air contaminants by passing ambient air through the air-purifying element.

Assigned protection factor (APF) [Reserved]

Atmosphere-supplying respirator means a respirator that supplies the respirator user with breathing air from a source independent of the ambient atmosphere, and includes supplied-air respirators (SARs) and self-contained breathing apparatus (SCBA) units.

Canister or cartridge means a container with a filter, sorbent, or catalyst, or combination of these items, which removes specific contaminants from the air passed through the container.

Demand respirator means an atmosphere-supplying respirator that admits breathing air to the facepiece only when a negative pressure is created inside the facepiece by inhalation.

Emergency situation means any occurrence such as, but not limited to, equipment failure, rupture of containers, or failure of control equipment that may or does result in an uncontrolled significant release of an airborne contaminant.

Employee exposure means exposure to a concentration of an airborne contaminant that would occur if the employee were not using respiratory protection.

End-of-service-life indicator (ESLI) means a system that warns the respirator user of the approach of the end of adequate respiratory protection, for example, that the sorbent is approaching saturation or is no longer effective.

Escape-only respirator means a respirator intended to be used only for emergency exit.

Filter or air purifying element means a component used in respirators to remove solid or liquid aerosols from the inspired air.

Filtering facepiece (dust mask) means a negative pressure particulate respirator with a filter as an integral part of the facepiece or with the entire facepiece composed of the filtering medium.

Fit factor means a quantitative estimate of the fit of a particular respirator to a specific individual, and typically estimates the ratio of the concentration of a substance in ambient air to its concentration inside the respirator when worn.

Fit test means the use of a protocol to qualitatively or quantitatively evaluate the fit of a respirator on an individual. See also qualitative fit test (QLFT) and quantitative fit test (QNFT).

Helmet means a rigid respiratory inlet covering that also provides head protection against impact and penetration.

High efficiency particulate air (HEPA) filter means a filter that is at least 99.97% efficient in removing monodispersed particles of 0.3 micrometers in diameter. The equivalent NIOSH 42 CFR 84 particulate filters are the N100, R100, and P100 filters.

Hood means a respiratory inlet covering that completely covers the head and neck and may also cover portions of the shoulders and torso.

Immediately dangerous to life or health (IDLH) means an atmosphere that poses an immediate threat to life, would cause irreversible adverse health effects, or would impair an individual's ability to escape from a dangerous atmosphere.

Interior structural firefighting means the physical activity of fire suppression, rescue or both, inside of buildings or enclosed structures which are involved in a fire situation beyond the incipient stage. (See 29 CFR 1910.155)

Loose-fitting facepiece means a respiratory inlet covering that is designed to form a partial seal with the face.

Maximum use concentration (MUC) [Reserved].

Negative pressure respirator (tight fitting) means a respirator in which the air pressure inside the facepiece is negative during inhalation with respect to the ambient air pressure outside the respirator.

Oxygen deficient atmosphere means an atmosphere with an oxygen content below 19.5% by volume.

Physician or other licensed health care professional (PLHCP) means an individual whose legally permitted scope of practice (i.e., license, registration, or certification) allows him or her to independently provide, or be delegated the responsibility to provide, some or all of the health care services required by paragraph (e) of this section.

Positive pressure respirator means a respirator in which the pressure inside the respiratory inlet covering exceeds the ambient air pressure outside the respirator.

Powered air-purifying respirator (PAPR) means an air-purifying respirator that uses a blower to force the ambient air through air-purifying elements to the inlet covering.

Pressure demand respirator means a positive pressure atmosphere-supplying respirator that admits breathing air to the facepiece when the positive pressure is reduced inside the facepiece by inhalation.

Qualitative fit test (QLFT) means a pass/fail fit test to assess the adequacy of respirator fit that relies on the individual's response to the test agent.

Quantitative fit test (QNFT) means an assessment of the adequacy of respirator fit by numerically measuring the amount of leakage into the respirator.

Respiratory inlet covering means that portion of a respirator that forms the protective barrier between the user's respiratory tract and an air-purifying device or breathing air source, or both. It may be a facepiece, helmet, hood, suit, or a mouthpiece respirator with nose clamp.

Self-contained breathing apparatus (SCBA) means an atmosphere-supplying respirator for which the breathing air source is designed to be carried by the user.

Service life means the period of time that a respirator, filter or sorbent, or other respiratory equipment provides adequate protection to the wearer.

Supplied-air respirator (SAR) or airline respirator means an atmosphere-supplying respirator for which the source of breathing air is not designed to be carried by the user.

This section means this respiratory protection standard.

Tight-fitting facepiece means a respiratory inlet covering that forms a complete seal with the face.

User seal check means an action conducted by the respirator user to determine if the respirator is properly seated to the face.

(c) *Respiratory protection program.* This paragraph requires the employer to develop and implement a written respiratory protection program with required worksite-specific procedures and elements for required respirator use. The program must be administered by a suitably trained program administrator. In addition, certain program elements may be required for voluntary use to prevent potential hazards associated with the use of the respirator. The Small Entity Compliance Guide contains criteria for the selection of a program administrator and a sample program that meets the requirements of this paragraph. Copies of the Small Entity Compliance Guide will be available on or about April 8, 1998 from the Occupational Safety and Health Administration's Office of Publications, Room N 3101, 200 Constitution Avenue, NW, Washington, DC 20210 (202-219-4667).

(c)(1) In any workplace where respirators are necessary to protect the health of the employee or whenever respirators are required by the employer, the employer shall establish and implement a written respiratory protection program with worksite-specific procedures. The program shall be updated as necessary to reflect those changes in workplace conditions that affect respirator use. The employer shall include in the program the following provisions of this section, as applicable:

(c)(1)(i) Procedures for selecting respirators for use in the workplace;

(c)(1)(ii) Medical evaluations of employees required to use respirators;

(c)(1)(iii) Fit testing procedures for tight-fitting respirators;

(c)(1)(iv) Procedures for proper use of respirators in routine and reasonably foreseeable emergency situations;

(c)(1)(v) Procedures and schedules for cleaning, disinfecting, storing, inspecting, repairing, discarding, and otherwise maintaining respirators;

(c)(1)(vi) Procedures to ensure adequate air quality, quantity, and flow of breathing air for atmosphere-supplying respirators;

(c)(1)(vii) Training of employees in the respiratory hazards to which they are potentially exposed during routine and emergency situations;

1910.134(c)(1)(viii)

(c)(1)(viii) Training of employees in the proper use of respirators, including putting on and removing them, any limitations on their use, and their maintenance; and

(c)(1)(ix) Procedures for regularly evaluating the effectiveness of the program.

(c)(2) Where respirator use is not required:

(c)(2)(i) An employer may provide respirators at the request of employees or permit employees to use their own respirators, if the employer determines that such respirator use will not in itself create a hazard. If the employer determines that any voluntary respirator use is permissible, the employer shall provide the respirator users with the information contained in Appendix D to this section ("Information for Employees Using Respirators When Not Required Under the Standard"); and

(c)(2)(ii) In addition, the employer must establish and implement those elements of a written respiratory protection program necessary to ensure that any employee using a respirator voluntarily is medically able to use that respirator, and that the respirator is cleaned, stored, and maintained so that its use does not present a health hazard to the user. Exception: Employers are not required to include in a written respiratory protection program those employees whose only use of respirators involves the voluntary use of filtering facepieces (dust masks).

(c)(3) The employer shall designate a program administrator who is qualified by appropriate training or experience that is commensurate with the complexity of the program to administer or oversee the respiratory protection program and conduct the required evaluations of program effectiveness.

(c)(4) The employer shall provide respirators, training, and medical evaluations at no cost to the employee.

(d) *Selection of respirators.* This paragraph requires the employer to evaluate respiratory hazard(s) in the workplace, identify relevant workplace and user factors, and base respirator selection on these factors. The paragraph also specifies appropriately protective respirators for use in IDLH atmospheres, and limits the selection and use of air-purifying respirators.

(d)(1) *General requirements.*

(d)(1)(i) The employer shall select and provide an appropriate respirator based on the respiratory hazard(s) to which the worker is exposed and workplace and user factors that affect respirator performance and reliability.

(d)(1)(ii) The employer shall select a NIOSH-certified respirator. The respirator shall be used in compliance with the conditions of its certification.

(d)(1)(iii) The employer shall identify and evaluate the respiratory hazard(s) in the workplace; this evaluation shall include a reasonable estimate of employee exposures to respiratory hazard(s) and an identification of the contaminant's chemical state and physical form. Where the employer cannot identify or reasonably estimate the employee exposure, the employer shall consider the atmosphere to be IDLH.

1910.134(d)(1)(iv)

(d)(1)(iv) The employer shall select respirators from a sufficient number of respirator models and sizes so that the respirator is acceptable to, and correctly fits, the user.

(d)(2) *Respirators for IDLH atmospheres.*

(d)(2)(i) The employer shall provide the following respirators for employee use in IDLH atmospheres:

Table II

Altitude (ft)	**Oxygen deficient atmospheres (% O_2) for which the employer may rely on atmosphere-supplying respirators**
Less than 3,001	16.0–19.5
3,001–4,000	16.4–19.5
4,001–5,000	17.1–19.5
5,001–6,000	17.8–19.5
6,001–7,000	18.5–19.5
7,001–8,000[1]	19.3–19.5

[1]Above 8,000 feet the exception does not apply. Oxygen-enriched breathing air must be supplied above 14,000 feet.

(d)(2)(i)(A) A full facepiece pressure demand SCBA certified by NIOSH for a minimum service life of thirty minutes, or

(d)(2)(i)(B) A combination full facepiece pressure demand supplied-air respirator (SAR) with auxiliary self-contained air supply.

(d)(2)(ii) Respirators provided only for escape from IDLH atmospheres shall be NIOSH-certified for escape from the atmosphere in which they will be used.

(d)(2)(iii) All oxygen-deficient atmospheres shall be considered IDLH. Exception: If the employer demonstrates that, under all foreseeable conditions, the oxygen concentration can be maintained within the ranges specified in Table II of this section (i.e., for the altitudes set out in the table), then any atmosphere-supplying respirator may be used.

(d)(3) *Respirators for atmospheres that are not IDLH.*

(d)(3)(i) The employer shall provide a respirator that is adequate to protect the health of the employee and ensure compliance with all other OSHA statutory and regulatory requirements, under routine and reasonably foreseeable emergency situations.

(d)(3)(i)(A) *Assigned Protection Factors (APFs)* [Reserved]

(d)(3)(i)(B) *Maximum Use Concentration (MUC)* [Reserved]

(d)(3)(ii) The respirator selected shall be appropriate for the chemical state and physical form of the contaminant.

(d)(3)(iii) For protection against gases and vapors, the employer shall provide:

(d)(3)(iii)(A) An atmosphere-supplying respirator, or

(d)(3)(iii)(B) An air-purifying respirator, provided that:

(d)(3)(iii)(B)(1) *(1)* The respirator is equipped with an end-of-service-life indicator (ESLI) certified by NIOSH for the contaminant; or

(d)(3)(iii)(B)(2) If there is no ESLI appropriate for conditions in the employer's workplace, the employer implements a change schedule for canisters and cartridges that is based on objective information or data that will ensure that canisters and cartridges are changed before the end of their service life. The employer shall describe in the respirator program the information and data relied upon and the basis for the canister and cartridge change schedule and the basis for reliance on the data.

(d)(3)(iv) For protection against particulates, the employer shall provide:

(d)(3)(iv)(A) An atmosphere-supplying respirator; or

(d)(3)(iv)(B) An air-purifying respirator equipped with a filter certified by NIOSH under 30 CFR part 11 as a high efficiency particulate air (HEPA) filter, or an air-purifying respirator equipped with a filter certified for particulates by NIOSH under 42 CFR part 84; or

(d)(3)(iv)(C) For contaminants consisting primarily of particles with mass median aerodynamic diameters (MMAD) of at least 2 micrometers, an air-purifying respirator equipped with any filter certified for particulates by NIOSH.

TABLE I. Assigned Protection Factors [Reserved]

1910.134(e)

(e) *Medical evaluation.* Using a respirator may place a physiological burden on employees that varies with the type of respirator worn, the job and workplace conditions in which the respirator is used, and the medical status of the employee. Accordingly, this paragraph specifies the minimum requirements for medical evaluation that employers must implement to determine the employee's ability to use a respirator.

(e)(1) *General.* The employer shall provide a medical evaluation to determine the employee's ability to use a respirator, before the employee is fit tested or required to use the respirator in the workplace. The employer may discontinue an employee's medical evaluations when the employee is no longer required to use a respirator.

(e)(2) *Medical evaluation procedures.*

(e)(2)(i) The employer shall identify a physician or other licensed health care professional (PLHCP) to perform medical evaluations using a medical questionnaire or an initial medical examination that obtains the same information as the medical questionnaire.

(e)(2)(ii) The medical evaluation shall obtain the information requested by the questionnaire in Sections 1 and 2, Part A of Appendix C of this section.

(e)(3) *Follow-up medical examination.*

(e)(3)(i) The employer shall ensure that a follow-up medical examination is provided for an employee who gives a positive response to any question among questions 1 through 8 in Section 2, Part A of Appendix C or whose initial medical examination demonstrates the need for a follow-up medical examination.

(e)(3)(ii) The follow-up medical examination shall include any medical tests, consultations, or diagnostic procedures that the PLHCP deems necessary to make a final determination.

(e)(4) *Administration of the medical questionnaire and examinations.*

(e)(4)(i) The medical questionnaire and examinations shall be administered confidentially during the employee's normal working hours or at a time and place convenient to the employee. The medical questionnaire shall be administered in a manner that ensures that the employee understands its content.

(e)(4)(ii) The employer shall provide the employee with an opportunity to discuss the questionnaire and examination results with the PLHCP.

(e)(5) *Supplemental information for the PLHCP.*

(e)(5)(i) The following information must be provided to the PLHCP before the PLHCP makes a recommendation concerning an employee's ability to use a respirator:

(e)(5)(i)(A) (A) The type and weight of the respirator to be used by the employee;

(e)(5)(i)(B) The duration and frequency of respirator use (including use for rescue and escape);

(e)(5)(i)(C) The expected physical work effort;

(e)(5)(i)(D) Additional protective clothing and equipment to be worn; and

(e)(5)(i)(E) Temperature and humidity extremes that may be encountered.

(e)(5)(ii) Any supplemental information provided previously to the PLHCP regarding an employee need not be provided for a subsequent medical evaluation if the information and the PLHCP remain the same.

(e)(5)(iii) The employer shall provide the PLHCP with a copy of the written respiratory protection program and a copy of this section.

Note to Paragraph (e)(5)(iii): When the employer replaces a PLHCP, the employer must ensure that the new PLHCP obtains this information, either by providing the documents directly to the PLHCP or having the documents transferred from the former PLHCP to the new PLHCP. However, OSHA does not expect employers to have employees medically reevaluated solely because a new PLHCP has been selected.

(e)(6) *Medical determination.* In determining the employee's ability to use a respirator, the employer shall:

(e)(6)(i) Obtain a written recommendation regarding the employee's ability to use the respirator from the PLHCP. The recommendation shall provide only the following information:

(e)(6)(i)(A) Any limitations on respirator use related to the medical condition of the employee, or relating to the workplace conditions in which the respirator will be used, including whether or not the employee is medically able to use the respirator;

(e)(6)(i)(B) The need, if any, for follow-up medical evaluations; and

(e)(6)(i)(C) A statement that the PLHCP has provided the employee with a copy of the PLHCP's written recommendation.

(e)(6)(ii) If the respirator is a negative pressure respirator and the PLHCP finds a medical condition that may place the employee's health at increased risk if the respirator is used, the employer shall provide a PAPR if the PLHCP's medical evaluation finds that the employee can use such a respirator; if a subsequent medical evaluation finds that the employee is medically able to use a negative pressure respirator, then the employer is no longer required to provide a PAPR.

(e)(7) *Additional medical evaluations.* At a minimum, the employer shall provide additional medical evaluations that comply with the requirements of this section if:

(e)(7)(i) An employee reports medical signs or symptoms that are related to ability to use a respirator;

(e)(7)(ii) A PLHCP, supervisor, or the respirator program administrator informs the employer that an employee needs to be reevaluated;

(e)(7)(iii) Information from the respiratory protection program, including observations made during fit testing and program evaluation, indicates a need for employee reevaluation; or

(e)(7)(iv) A change occurs in workplace conditions (e.g., physical work effort, protective clothing, temperature) that may result in a substantial increase in the physiological burden placed on an employee.

(f) *Fit testing.* This paragraph requires that, before an employee may be required to use any respirator with a negative or positive pressure tight-fitting facepiece, the employee must be fit tested with the same make, model, style, and size of respirator that will be used. This paragraph specifies the kinds of fit tests allowed, the procedures for conducting them, and how the results of the fit tests must be used.

(f)(1) The employer shall ensure that employees using a tight-fitting facepiece respirator pass an appropriate qualitative fit test (QLFT) or quantitative fit test (QNFT) as stated in this paragraph.

(f)(2) The employer shall ensure that an employee using a tight-fitting facepiece respirator is fit tested prior to initial use of the respirator, whenever a different respirator facepiece (size, style, model or make) is used, and at least annually thereafter.

(f)(3) The employer shall conduct an additional fit test whenever the employee reports, or the employer, PLHCP, supervisor, or program administrator makes visual observations of, changes in the employee's physical condition that could affect respirator fit. Such conditions include, but are not limited to, facial scarring, dental changes, cosmetic surgery, or an obvious change in body weight.

(f)(4) If after passing a QLFT or QNFT, the employee subsequently notifies the employer, program administrator, supervisor, or PLHCP that the fit of the respirator is unacceptable, the employee shall be given a reasonable opportunity to select a different respirator facepiece and to be retested.

1910.134(f)(5)

(f)(5) The fit test shall be administered using an OSHA-accepted QLFT or QNFT protocol. The OSHA-accepted QLFT and QNFT protocols and procedures are contained in Appendix A of this section.

(f)(6) QLFT may only be used to fit test negative pressure air-purifying respirators that must achieve a fit factor of 100 or less.

(f)(7) If the fit factor, as determined through an OSHA-accepted QNFT protocol, is equal to or greater than 100 for tight-fitting half facepieces, or equal to or greater than 500 for tight-fitting full facepieces, the QNFT has been passed with that respirator.

(f)(8) Fit testing of tight-fitting atmosphere-supplying respirators and tight-fitting powered air-purifying respirators shall be accomplished by performing quantitative or qualitative fit testing in the negative pressure mode, regardless of the mode of operation (negative or positive pressure) that is used for respiratory protection.

(f)(1)(8)(i) Qualitative fit testing of these respirators shall be accomplished by temporarily converting the respirator user's actual facepiece into a negative pressure respirator with appropriate filters, or by using an identical negative pressure air-purifying respirator facepiece with the same sealing surfaces as a surrogate for the atmosphere-supplying or powered air-purifying respirator facepiece.

(f)(1)(8)(ii) Quantitative fit testing of these respirators shall be accomplished by modifying the facepiece to allow sampling inside the facepiece in the breathing zone of the user, midway between the nose and mouth. This requirement shall be accomplished by installing a permanent sampling probe onto a surrogate facepiece, or by using a sampling adapter designed to temporarily provide a means of sampling air from inside the facepiece.

(f)(1)(8)(iii) Any modifications to the respirator facepiece for fit testing shall be completely removed, and the facepiece restored to NIOSH-approved configuration, before that facepiece can be used in the workplace.

(g) *Use of respirators.* This paragraph requires employers to establish and implement procedures for the proper use of respirators. These requirements include prohibiting conditions that may result in facepiece seal leakage, preventing employees from removing respirators in hazardous environments, taking actions to ensure continued effective respirator operation throughout the work shift, and establishing procedures for the use of respirators in IDLH atmospheres or in interior structural firefighting situations.

(g)(1) *Facepiece seal protection.*

(g)(1)(i) The employer shall not permit respirators with tight-fitting facepieces to be worn by employees who have:

(g)(1)(i)(A) Facial hair that comes between the sealing surface of the facepiece and the face or that interferes with valve function; or

(g)(1)(i)(B) Any condition that interferes with the face-to-facepiece seal or valve function.

(g)(1)(ii) If an employee wears corrective glasses or goggles or other personal protective equipment, the employer shall ensure that such equipment is worn in a manner that does not interfere with the seal of the facepiece to the face of the user.

(g)(1)(iii) For all tight-fitting respirators, the employer shall ensure that employees perform a user seal check each time they put on the respirator using the procedures in Appendix B-1 or procedures recommended by the respirator manufacturer that the employer demonstrates are as effective as those in Appendix B-1 of this section.

(g)(2) *Continuing respirator effectiveness.*

(g)(2)(i) Appropriate surveillance shall be maintained of work area conditions and degree of employee exposure or stress. When there is a change in work area conditions or degree of employee exposure or stress that may affect respirator effectiveness, the employer shall reevaluate the continued effectiveness of the respirator.

(g)(2)(ii) The employer shall ensure that employees leave the respirator use area:

1910.134(g)(2)(ii)(A)

(g)(2)(ii)(A) To wash their faces and respirator facepieces as necessary to prevent eye or skin irritation associated with respirator use; or

(g)(2)(ii)(B) If they detect vapor or gas breakthrough, changes in breathing resistance, or leakage of the facepiece; or

(g)(2)(ii)(C) To replace the respirator or the filter, cartridge, or canister elements.

(g)(2)(iii) If the employee detects vapor or gas breakthrough, changes in breathing resistance, or leakage of the facepiece, the employer must replace or repair the respirator before allowing the employee to return to the work area.

(g)(3) *Procedures for IDLH atmospheres.* For all IDLH atmospheres, the employer shall ensure that:

(g)(3)(i) One employee or, when needed, more than one employee is located outside the IDLH atmosphere;

(g)(3)(ii) Visual, voice, or signal line communication is maintained between the employee(s) in the IDLH atmosphere and the employee(s) located outside the IDLH atmosphere;

(g)(3)(iii) The employee(s) located outside the IDLH atmosphere are trained and equipped to provide effective emergency rescue;

(g)(3)(iv) The employer or designee is notified before the employee(s) located outside the IDLH atmosphere enter the IDLH atmosphere to provide emergency rescue;

(g)(3)(v) The employer or designee authorized to do so by the employer, once notified, provides necessary assistance appropriate to the situation;

(g)(3)(vi) Employee(s) located outside the IDLH atmospheres are equipped with:

(g)(3)(vi)(A) Pressure demand or other positive pressure SCBAs, or a pressure demand or other positive pressure supplied-air respirator with auxiliary SCBA; and either

(g)(3)(vi)(B) Appropriate retrieval equipment for removing the employee(s) who enter(s) these hazardous atmospheres where retrieval equipment would contribute to the rescue of the employee(s) and would not increase the overall risk resulting from entry; or

(g)(3)(vi)(C) Equivalent means for rescue where retrieval equipment is not required under paragraph (g)(3)(vi)(B).

(g)(4) *Procedures for interior structural firefighting.* In addition to the requirements set forth under paragraph (g)(3), in interior structural fires, the employer shall ensure that:

(g)(4)(i) At least two employees enter the IDLH atmosphere and remain in visual or voice contact with one another at all times;

(g)(4)(ii) At least two employees are located outside the IDLH atmosphere; and

(g)(4)(iii) All employees engaged in interior structural firefighting use SCBAs.

Note 1 to paragraph (g): One of the two individuals located outside the IDLH atmosphere may be assigned to an additional role, such as incident commander in charge of the emergency or safety officer, so long as this individual is able to perform assistance or rescue activities without jeopardizing the safety or health of any firefighter working at the incident.

Note 2 to paragraph (g): Nothing in this section is meant to preclude firefighters from performing emergency rescue activities before an entire team has assembled.

(h) *Maintenance and care of respirators.* This paragraph requires the employer to provide for the cleaning and disinfecting, storage, inspection, and repair of respirators used by employees.

(h)(1) *Cleaning and disinfecting.* The employer shall provide each respirator user with a respirator that is clean, sanitary, and in good working order. The employer shall ensure that respirators are cleaned and disinfected using the procedures in Appendix B-2 of this section, or procedures recommended by the respirator manufacturer, provided that such procedures are of equivalent effectiveness. The respirators shall be cleaned and disinfected at the following intervals:

(h)(1)(i) Respirators issued for the exclusive use of an employee shall be cleaned and disinfected as often as necessary to be maintained in a sanitary condition;

(h)(1)(ii) Respirators issued to more than one employee shall be cleaned and disinfected before being worn by different individuals;

(h)(1)(iii) Respirators maintained for emergency use shall be cleaned and disinfected after each use; and

(h)(1)(iv) Respirators used in fit testing and training shall be cleaned and disinfected after each use.

(h)(2) *Storage.* The employer shall ensure that respirators are stored as follows:

(h)(2)(i) All respirators shall be stored to protect them from damage, contamination, dust, sunlight, extreme temperatures, excessive moisture, and damaging chemicals, and they shall be packed or stored to prevent deformation of the facepiece and exhalation valve.

(h)(2)(ii) In addition to the requirements of paragraph (h)(2)(i) of this section, emergency respirators shall be:

(h)(2)(ii)(A) Kept accessible to the work area;

(h)(2)(ii)(B) Stored in compartments or in covers that are clearly marked as containing emergency respirators; and

(h)(2)(ii)(C) Stored in accordance with any applicable manufacturer instructions.

1910.134(h)(3)

(h)(3) *Inspection.*

(h)(3)(i) The employer shall ensure that respirators are inspected as follows:

(h)(3)(i)(A) All respirators used in routine situations shall be inspected before each use and during cleaning;

(h)(3)(i)(B) All respirators maintained for use in emergency situations shall be inspected at least monthly and in accordance with the manufacturer's recommendations, and shall be checked for proper function before and after each use; and

(h)(3)(i)(C) Emergency escape-only respirators shall be inspected before being carried into the workplace for use.

(h)(3)(ii) The employer shall ensure that respirator inspections include the following:

(h)(3)(ii)(A) A check of respirator function, tightness of connections, and the condition of the various parts including, but not limited to, the facepiece, head straps, valves, connecting tube, and cartridges, canisters or filters; and

(h)(3)(ii)(B) A check of elastomeric parts for pliability and signs of deterioration.

(h)(3)(iii) In addition to the requirements of paragraphs (h)(3)(i) and (ii) of this section, self-contained breathing apparatus shall be inspected monthly. Air and oxygen cylinders shall be maintained in a fully charged state and shall be recharged when the pressure falls to 90% of the manufacturer's recommended pressure level. The employer shall determine that the regulator and warning devices function properly.

(h)(3)(iv) For respirators maintained for emergency use, the employer shall:

(h)(3)(iv)(A) Certify the respirator by documenting the date the inspection was performed, the name (or signature) of the person who made the inspection, the findings, required remedial action, and a serial number or other means of identifying the inspected respirator; and (B) Provide this information on a tag or label that is attached to the storage compartment for the respirator, is kept with the respirator, or is included in inspection reports stored as paper or electronic files. This information shall be maintained until replaced following a subsequent certification.

(h)(4) *Repairs.* The employer shall ensure that respirators that fail an inspection or are otherwise found to be defective are removed from service, and are discarded or repaired or adjusted in accordance with the following procedures:

(h)(4)(i) Repairs or adjustments to respirators are to be made only by persons appropriately trained to perform such operations and shall use only the respirator manufacturer's NIOSH-approved parts designed for the respirator;

(h)(4)(ii) Repairs shall be made according to the manufacturer's recommendations and specifications for the type and extent of repairs to be performed; and

(h)(4)(iii) Reducing and admission valves, regulators, and alarms shall be adjusted or repaired only by the manufacturer or a technician trained by the manufacturer.

(i) *Breathing air quality and use.* This paragraph requires the employer to provide employees using atmosphere-supplying respirators (supplied-air and SCBA) with breathing gases of high purity.

(i)(1) The employer shall ensure that compressed air, compressed oxygen, liquid air, and liquid oxygen used for respiration accords with the following specifications:

(i)(1)(i) Compressed and liquid oxygen shall meet the United States Pharmacopoeia requirements for medical or breathing oxygen; and

1910.134(i)(1)(ii)

(i)(1)(ii) Compressed breathing air shall meet at least the requirements for Grade D breathing air described in ANSI/Compressed Gas Association Commodity Specification for Air, G-7.1-1989, to include:

(i)(1)(ii)(A) Oxygen content (v/v) of 19.5-23.5%;

(i)(1)(ii)(B) Hydrocarbon (condensed) content of 5 milligrams per cubic meter of air or less;

(i)(1)(ii)(C) Carbon monoxide (CO) content of 10 ppm or less;

(i)(1)(ii)(D) Carbon dioxide content of 1,000 ppm or less; and

(i)(1)(ii)(E) Lack of noticeable odor.

(i)(2) The employer shall ensure that compressed oxygen is not used in atmosphere-supplying respirators that have previously used compressed air.

(i)(3) The employer shall ensure that oxygen concentrations greater than 23.5% are used only in equipment designed for oxygen service or distribution.

(i)(4) The employer shall ensure that cylinders used to supply breathing air to respirators meet the following requirements:

(i)(4)(i) Cylinders are tested and maintained as prescribed in the Shipping Container Specification Regulations of the Department of Transportation (49 CFR part 173 and part 178);

(i)(4)(ii) Cylinders of purchased breathing air have a certificate of analysis from the supplier that the breathing air meets the requirements for Grade D breathing air; and

(i)(4)(iii) The moisture content in the cylinder does not exceed a dew point of –50°F (–45.6°C) at 1 atmosphere pressure.

(i)(5) The employer shall ensure that compressors used to supply breathing air to respirators are constructed and situated so as to:

(i)(5)(i) Prevent entry of contaminated air into the air-supply system;

(i)(5)(ii) Minimize moisture content so that the dew point at 1 atmosphere pressure is 10°F (5.56 °C) below the ambient temperature;

(i)(5)(iii) Have suitable in-line air-purifying sorbent beds and filters to further ensure breathing air quality. Sorbent beds and filters shall be maintained and replaced or refurbished periodically following the manufacturer's instructions.

(i)(5)(iv) Have a tag containing the most recent change date and the signature of the person authorized by the employer to perform the change. The tag shall be maintained at the compressor.

(i)(6) For compressors that are not oil-lubricated, the employer shall ensure that carbon monoxide levels in the breathing air do not exceed 10 ppm.

(i)(7) For oil-lubricated compressors, the employer shall use a high-temperature or carbon monoxide alarm, or both, to monitor carbon monoxide levels. If only high-temperature alarms are used, the air supply shall be monitored at intervals sufficient to prevent carbon monoxide in the breathing air from exceeding 10 ppm.

(i)(8) The employer shall ensure that breathing air couplings are incompatible with outlets for nonrespirable worksite air or other gas systems. No asphyxiating substance shall be introduced into breathing air lines.

(i)(9) The employer shall use breathing gas containers marked in accordance with the NIOSH respirator certification standard, 42 CFR part 84.

(j) *Identification of filters, cartridges, and canisters.* The employer shall ensure that all filters, cartridges and canisters used in the workplace are labeled and color coded with the NIOSH approval label and that the label is not removed and remains legible.

(k) *Training and information.* This paragraph requires the employer to provide effective training to employees who are required to use respirators. The training must be comprehensive, understandable, and recur annually, and more often if necessary. This paragraph also requires the employer to provide the basic information on respirators in Appendix D of this section to employees who wear respirators when not required by this section or by the employer to do so.

(k)(1) The employer shall ensure that each employee can demonstrate knowledge of at least the following:

1910.134(k)(1)(i)

(k)(1)(i) Why the respirator is necessary and how improper fit, usage, or maintenance can compromise the protective effect of the respirator;

(k)(1)(ii) What the limitations and capabilities of the respirator are;

(k)(1)(iii) How to use the respirator effectively in emergency situations, including situations in which the respirator malfunctions;

(k)(1)(iv) How to inspect, put on and remove, use, and check the seals of the respirator;

(k)(1)(v) What the procedures are for maintenance and storage of the respirator;

(k)(1)(vi) How to recognize medical signs and symptoms that may limit or prevent the effective use of respirators; and

(k)(1)(vii) The general requirements of this section.

(k)(2) The training shall be conducted in a manner that is understandable to the employee.

(k)(3) The employer shall provide the training prior to requiring the employee to use a respirator in the workplace.

(k)(4) An employer who is able to demonstrate that a new employee has received training within the last 12 months that addresses the elements specified in paragraph (k)(1)(i) through (vii) is not required to repeat such training provided that, as required by paragraph (k)(1), the employee can demonstrate knowledge of those element(s). Previous training not repeated initially by the employer must be provided no later than 12 months from the date of the previous training.

(k)(5) Retraining shall be administered annually, and when the following situations occur:

(k)(5)(i) Changes in the workplace or the type of respirator render previous training obsolete;

(k)(5)(ii) Inadequacies in the employee's knowledge or use of the respirator indicate that the employee has not retained the requisite understanding or skill; or

(k)(5)(iii) Any other situation arises in which retraining appears necessary to ensure safe respirator use.

(k)(6) The basic advisory information on respirators, as presented in Appendix D of this section, shall be provided by the employer in any written or oral format, to employees who wear respirators when such use is not required by this section or by the employer.

(l) *Program evaluation.* This section requires the employer to conduct evaluations of the workplace to ensure that the written respiratory protection program is being properly implemented, and to consult employees to ensure that they are using the respirators properly.

(l)(1) The employer shall conduct evaluations of the workplace as necessary to ensure that the provisions of the current written program are being effectively implemented and that it continues to be effective.

(l)(2) The employer shall regularly consult employees required to use respirators to assess the employees' views on program effectiveness and to identify any problems. Any problems that are identified during this assessment shall be corrected. Factors to be assessed include, but are not limited to:

1910.134((l)(2)(i)

(l)(2)(i) Respirator fit (including the ability to use the respirator without interfering with effective workplace performance);

(l)(2)(ii) Appropriate respirator selection for the hazards to which the employee is exposed;

(l)(2)(iii) Proper respirator use under the workplace conditions the employee encounters; and

(l)(2)(iv) Proper respirator maintenance.

(m) *Recordkeeping.* This section requires the employer to establish and retain written information regarding medical evaluations, fit testing, and the respirator program. This information will facilitate employee involvement in the respirator program, assist the employer in auditing the adequacy of the program, and provide a record for compliance determinations by OSHA.

1910.134(m)(1)

(m)(1) *Medical evaluation.* Records of medical evaluations required by this section must be retained and made available in accordance with 29 CFR 1910.1020.

(m)(2) *Fit testing.*

(m)(2)(i) The employer shall establish a record of the qualitative and quantitative fit tests administered to an employee including:

(m)(2)(i)(A) The name or identification of the employee tested;

(m)(2)(i)(B) Type of fit test performed;

(m)(2)(i)(C) Specific make, model, style, and size of respirator tested;

(m)(2)(i)(D) Date of test; and

(m)(2)(i)(E) The pass/fail results for QLFTs or the fit factor and strip chart recording or other recording of the test results for QNFTs.

(m)(2)(ii) Fit test records shall be retained for respirator users until the next fit test is administered.

(m)(3) A written copy of the current respirator program shall be retained by the employer.

(m)(4) Written materials required to be retained under this paragraph shall be made available upon request to affected employees and to the Assistant Secretary or designee for examination and copying.

(n) *Dates.*

(n)(1) *Effective date.* This section is effective April 8, 1998. The obligations imposed by this section commence on the effective date unless otherwise noted in this paragraph. Compliance with obligations that do not commence on the effective date shall occur no later than the applicable start-up date.

(n)(2) *Compliance dates.* All obligations of this section commence on the effective date except as follows:

1910.134(n)(2)(i)

(n)(2)(i) The determination that respirator use is required (paragraph (a)) shall be completed no later than September 8, 1998.

(n)(2)(ii) Compliance with provisions of this section for all other provisions shall be completed no later than October 5, 1998.

Appendix 1A

Appendix A to § 1910.134 (Mandatory): Fit Testing Procedures

Part I. OSHA—Accepted Fit Test Protocols

A. Fit Testing Procedures—General Requirements

The employer shall conduct fit testing using the following procedures. The requirements in this appendix apply to all OSHA-accepted fit test methods, both QLFT and QNFT.

1. The test subject shall be allowed to pick the most acceptable respirator from a sufficient number of respirator models and sizes so that the respirator is acceptable to, and correctly fits, the user.

2. Prior to the selection process, the test subject shall be shown how to put on a respirator, how it should be positioned on the face, how to set strap tension and how to determine an acceptable fit. A mirror shall be available to assist the subject in evaluating the fit and positioning of the respirator. This instruction may not constitute the subject's formal training on respirator use, because it is only a review.

3. The test subject shall be informed that he/she is being asked to select the respirator that provides the most acceptable fit. Each respirator represents a different size and shape, and if fitted and used properly, will provide adequate protection.

4. The test subject shall be instructed to hold each chosen facepiece up to the face and eliminate those that obviously do not give an acceptable fit.

5. The more acceptable facepieces are noted in case the one selected proves unacceptable; the most comfortable mask is donned and worn at least five minutes to assess comfort. Assistance in assessing comfort can be given by discussing the points in the following item A.6. If the test subject is not familiar with using a particular respirator, the test

subject shall be directed to don the mask several times and to adjust the straps each time to become adept at setting proper tension on the straps.

6. Assessment of comfort shall include a review of the following points with the test subject and allowing the test subject adequate time to determine the comfort of the respirator:

(a) Position of the mask on the nose

(b) Room for eye protection

(c) Room to talk

(d) Position of mask on face and cheeks

7. The following criteria shall be used to help determine the adequacy of the respirator fit:

(a) Chin properly placed;

(b)Adequate strap tension, not overly tightened;

(c) Fit across nose bridge;

(d) Respirator of proper size to span distance from nose to chin;

(e) Tendency of respirator to slip;

(f) Self-observation in mirror to evaluate fit and respirator position.

8. The test subject shall conduct a user seal check, either the negative and positive pressure seal checks described in Appendix B1 of this section or those recommended by the respirator manufacturer which provide equivalent protection to the procedures in Appendix B1. Before conducting the negative and positive pressure checks, the subject shall be told to seat the mask on the face by moving the head from side-to-side and up and down slowly while taking in a few slow deep breaths. Another facepiece shall be selected and retested if the test subject fails the user seal check tests.

9. The test shall not be conducted if there is any hair growth between the skin and the facepiece sealing surface, such as stubble beard growth, beard, mustache or sideburns which cross the respirator sealing surface. Any type of apparel which interferes with a satisfactory fit shall be altered or removed.

10. If a test subject exhibits difficulty in breathing during the tests, she or he shall be referred to a physician or other licensed health care professional, as appropriate, to determine whether the test subject can wear a respirator while performing her or his duties.

11. If the employee finds the fit of the respirator unacceptable, the test subject shall be given the opportunity to select a different respirator and to be retested.

12. Exercise regimen. Prior to the commencement of the fit test, the test subject shall be given a description of the fit test and the test subject's responsibilities during the test procedure. The description of the process shall include a description of the test exercises that the subject will be performing. The respirator to be tested shall be worn for at least 5 minutes before the start of the fit test.

13. The fit test shall be performed while the test subject is wearing any applicable safety equipment that may be worn during actual respirator use which could interfere with respirator fit.

14. Test Exercises. (a) The following test exercises are to be performed for all fit testing methods prescribed in this appendix, except for the CNP method. A separate fit testing exercise regimen is contained in the CNP protocol. The test subject shall perform exercises, in the test environment, in the following manner:

(1) Normal breathing. In a normal standing position, without talking, the subject shall breathe normally.

(2) Deep breathing. In a normal standing position, the subject shall breathe slowly and deeply, taking caution so as not to hyperventilate.

(3) Turning head side to side. Standing in place, the subject shall slowly turn his/her head from side to side between the extreme positions on each side. The head shall be held at each extreme momentarily so the subject can inhale at each side.

(4) Moving head up and down. Standing in place, the subject shall slowly move his/her head up and down. The subject shall be instructed to inhale in the up position (i.e., when looking toward the ceiling).

(5) Talking. The subject shall talk out loud slowly and loud enough so as to be heard clearly by the test conductor. The subject can read from a prepared text such as the Rainbow Passage, count backward from 100, or recite a memorized poem or song.

Rainbow Passage

When the sunlight strikes raindrops in the air, they act like a prism and form a rainbow. The rainbow is a division of white light into many beautiful colors. These take the shape of a long round arch, with its path high above, and its two ends apparently beyond the horizon. There is, according to legend, a boiling pot of gold at one end. People look, but no one ever finds it. When a man looks for something beyond reach, his friends say he is looking for the pot of gold at the end of the rainbow.

(6) Grimace. The test subject shall grimace by smiling or frowning. (This applies only to QNFT testing; it is not performed for QLFT)

(7) Bending over. The test subject shall bend at the waist as if he/she were to touch his/her toes. Jogging in place shall be substituted for this exercise in those test environments such as shroud type QNFT or QLFT units that do not permit bending over at the waist. (8) Normal breathing. Same as exercise (1).

(b)Each test exercise shall be performed for one minute except for the grimace exercise which shall be performed for 15 seconds. The test subject shall be questioned by the test conductor regarding the comfort of the respirator upon completion of the protocol. If it has become unacceptable, another model of respirator shall be tried. The respirator shall not be adjusted once the fit test exercises begin. Any adjustment voids the test, and the fit test must be repeated.

B. Qualitative Fit Test (QLFT) Protocols

1. General

(a) The employer shall ensure that persons administering QLFT are able to prepare test solutions, calibrate equipment and perform tests properly, recognize invalid tests, and ensure that test equipment is in proper working order.

(b) The employer shall ensure that QLFT equipment is kept clean and well maintained so as to operate within the parameters for which it was designed.

2. Isoamyl Acetate Protocol

Note: This protocol is not appropriate to use for the fit testing of particulate respirators. If used to fit test particulate respirators, the respirator must be equipped with an organic vapor filter.

(a) Odor Threshold Screening

Odor threshold screening, performed without wearing a respirator, is intended to determine if the individual tested can detect the odor of isoamyl acetate at low levels.

(1) Three 1 liter glass jars with metal lids are required.

(2) Odor-free water (e.g., distilled or spring water) at approximately 25°C (77°F) shall be used for the solutions.

(3) The isoamyl acetate (IAA) (also known at isopentyl acetate) stock solution is prepared by adding 1 ml of pure IAA to 800 ml of odor-free water in a 1 liter jar, closing the lid and shaking for 30 seconds. A new solution shall be prepared at least weekly.

(4) The screening test shall be conducted in a room separate from the room used for actual fit testing. The two rooms shall be well-ventilated to prevent the odor of IAA from becoming evident in the general room air where testing takes place.

(5) The odor test solution is prepared in a second jar by placing 0.4 ml of the stock solution into 500 ml of odor-free water using a clean dropper or pipette. The solution shall be shaken for 30 seconds and allowed to stand for two to three minutes so that the IAA concentration above the liquid may reach equilibrium. This solution shall be used for only one day.

(6) A test blank shall be prepared in a third jar by adding 500 cc of odor-free water.

(7) The odor test and test blank jar lids shall be labeled (e.g., 1 and 2) for jar identification. Labels shall be placed on the lids so that they can be peeled off periodically and switched to maintain the integrity of the test.

(8) The following instruction shall be typed on a card and placed on the table in front of the two test jars (i.e., 1 and 2): "The purpose of this test is to determine if you can smell banana oil at a low concentration. The two bottles in front of you contain water. One of these bottles also contains a small amount of banana oil. Be sure the covers are on tight, then shake each bottle for two seconds. Unscrew the lid of each bottle, one at a time, and sniff at the mouth of the bottle. Indicate to the test conductor which bottle contains banana oil."

(9) The mixtures used in the IAA odor detection test shall be prepared in an area separate from where the test is performed, in order to prevent olfactory fatigue in the subject.

(10) If the test subject is unable to correctly identify the jar containing the odor test solution, the IAA qualitative fit test shall not be performed.

(11) If the test subject correctly identifies the jar containing the odor test solution, the test subject may proceed to respirator selection and fit testing.

(b) Isoamyl Acetate Fit Test

(1) The fit test chamber shall be a clear 55-gallon drum liner suspended inverted over a 2-foot diameter frame so that the top of the chamber is about 6 inches above the test subject's head. If no drum liner is available, a similar chamber shall be constructed using plastic sheeting. The inside top center of the chamber shall have a small hook attached.

(2) Each respirator used for the fitting and fit testing shall be equipped with organic vapor cartridges or offer protection against organic vapors.

(3) After selecting, donning, and properly adjusting a respirator, the test subject shall wear it to the fit testing room. This room shall be separate from the room used for odor threshold screening and respirator selection, and shall be well-ventilated, as by an exhaust fan or lab hood, to prevent general room contamination.

(4) A copy of the test exercises and any prepared text from which the subject is to read shall be taped to the inside of the test chamber.

(5) Upon entering the test chamber, the test subject shall be given a 6-inch by 5-inch piece of paper towel, or other porous, absorbent, single-ply material, folded in half and wetted with 0.75 ml of pure IAA. The test subject shall hang the wet towel on the hook at the top of the chamber. An IAA test swab or ampule may be substituted for the IAA wetted paper towel provided it has been demonstrated that the alternative IAA source will generate an IAA test atmosphere with a concentration equivalent to that generated by the paper towel method.

(6) Allow two minutes for the IAA test concentration to stabilize before starting the fit test exercises. This would be an appropriate time to talk with the test subject; to explain the fit test, the importance of his/her cooperation, and the purpose for the test exercises; or to demonstrate some of the exercises.

(7) If at any time during the test, the subject detects the banana-like odor of IAA, the test is failed. The subject shall quickly exit from the test chamber and leave the test area to avoid olfactory fatigue.

(8) If the test is failed, the subject shall return to the selection room and remove the respirator. The test subject shall repeat the odor sensitivity test, select and put on another respirator, return to the test area and again begin the fit test procedure described in (b) (1) through (7) above. The process continues until a respirator that fits well has been found. Should the odor sensitivity test be failed, the subject shall wait at least 5 minutes before retesting. Odor sensitivity will usually have returned by this time.

(9) If the subject passes the test, the efficiency of the test procedure shall be demonstrated by having the subject break the respirator face seal and take a breath before exiting the chamber.

(10) When the test subject leaves the chamber, the subject shall remove the saturated towel and return it to the person conducting the test, so that there is no significant IAA concentration buildup in the chamber during subsequent tests. The used towels shall be kept in a self-sealing plastic bag to keep the test area from being contaminated.

3. Saccharin Solution Aerosol Protocol

The entire screening and testing procedure shall be explained to the test subject prior to the conduct of the screening test.

(a) Taste threshold screening. The saccharin taste threshold screening, performed without wearing a respirator, is intended to determine whether the individual being tested can detect the taste of saccharin.

(1) During threshold screening as well as during fit testing, subjects shall wear an enclosure about the head and shoulders that is approximately 12 inches in diameter by 14 inches tall with at least the front portion clear and that allows free movements of the head when a respirator is worn. An enclosure substantially similar to the 3M hood assembly, parts # FT 14 and # FT 15 combined, is adequate.

(2) The test enclosure shall have a 3/4-inch (1.9-cm) hole in front of the test subject's nose and mouth area to accommodate the nebulizer nozzle.

(3) The test subject shall don the test enclosure. Throughout the threshold screening test, the test subject shall breathe through his/her slightly open mouth with tongue extended. The subject is instructed to report when he/she detects a sweet taste.

(4) Using a DeVilbiss Model 40 Inhalation Medication Nebulizer or equivalent, the test conductor shall spray the threshold check solution into the enclosure. The nozzle is directed away from the nose and mouth of the person. This nebulizer shall be clearly marked to distinguish it from the fit test solution nebulizer..

(5) The threshold check solution is prepared by dissolving 0.83 gram of sodium saccharin USP in 100 ml of warm water. It can be prepared by putting 1 ml of the fit test solution (see (b)(5) below) in 100 ml of distilled water.

(6) To produce the aerosol, the nebulizer bulb is firmly squeezed so that it collapses completely, then released and allowed to fully expand.

(7) Ten squeezes are repeated rapidly and then the test subject is asked whether the saccharin can be tasted. If the test subject reports tasting the sweet taste during the ten squeezes, the screening test is completed. The taste threshold is noted as ten regardless of the number of squeezes actually completed.

(8) If the first response is negative, ten more squeezes are repeated rapidly and the test subject is again asked whether the saccharin is tasted. If the test subject reports tasting the sweet taste during the second ten squeezes, the screening test is completed. The taste threshold is noted as twenty regardless of the number of squeezes actually completed.

(9) If the second response is negative, ten more squeezes are repeated rapidly and the test subject is again asked whether the saccharin is tasted. If the test subject reports tasting the sweet taste during the third set of ten squeezes, the screening test is completed. The taste threshold is noted as thirty regardless of the number of squeezes actually completed.

(10) The test conductor will take note of the number of squeezes required to solicit a taste response.

(11) If the saccharin is not tasted after 30 squeezes (step 10), the test subject is unable to taste saccharin and may not perform the saccharin fit test.

Note to paragraph 3. (a): If the test subject eats or drinks something sweet before the screening test, he/she may be unable to taste the weak saccharin solution.

(12) If a taste response is elicited, the test subject shall be asked to take note of the taste for reference in the fit test.

(13) Correct use of the nebulizer means that approximately 1 ml of liquid is used at a time in the nebulizer body.

(14) The nebulizer shall be thoroughly rinsed in water, shaken dry, and refilled at least each morning and afternoon or at least every four hours.

(b) Saccharin solution aerosol fit test procedure.

(1) The test subject may not eat, drink (except plain water), smoke, or chew gum for 15 minutes before the test.

(2) The fit test uses the same enclosure described in 3. (a) above.

(3) The test subject shall don the enclosure while wearing the respirator selected in section I. A. of this appendix. The respirator shall be properly adjusted and equipped with a particulate filter(s).

(4) A second DeVilbiss Model 40 Inhalation Medication Nebulizer or equivalent is used to spray the fit test solution into the enclosure. This nebulizer shall be clearly marked to distinguish it from the screening test solution nebulizer.

(5) The fit test solution is prepared by adding 83 grams of sodium saccharin to 100 ml of warm water.

(6) As before, the test subject shall breathe through the slightly open mouth with tongue extended, and report if he/she tastes the sweet taste of saccharin.

(7) The nebulizer is inserted into the hole in the front of the enclosure and an initial concentration of saccharin fit test solution is sprayed into the enclosure using the same number of squeezes (either 10, 20 or 30 squeezes) based on the number of squeezes required to elicit a taste response as noted during the screening test. A minimum of 10 squeezes is required.

(8) After generating the aerosol, the test subject shall be instructed to perform the exercises in section I.A.14. of this appendix.

(9) Every 30 seconds the aerosol concentration shall be replenished using one half the original number of squeezes used initially (e.g., 5, 10 or 15).

(10) The test subject shall indicate to the test conductor if at any time during the fit test the taste of saccharin is detected. If the test subject does not report tasting the saccharin, the test is passed.

(11) If the taste of saccharin is detected, the fit is deemed unsatisfactory and the test is failed. A different respirator shall be tried and the entire test procedure is repeated (taste threshold screening and fit testing).

(12) Since the nebulizer has a tendency to clog during use, the test operator must make periodic checks of the nebulizer to ensure that it is not clogged. If clogging is found at the end of the test session, the test is invalid.

4. Bitrex™ (Denatonium Benzoate) Solution Aerosol Qualitative Fit Test Protocol

The Bitrex™ (denatonium benzoate) solution aerosol QLFT protocol uses the published saccharin test protocol because that protocol is widely accepted. Bitrex is routinely used as a taste aversion agent in household liquids which children should not be drinking and is endorsed by the American Medical Association, the National Safety Council, and the American Association of Poison Control Centers. The entire screening and testing procedure shall be explained to the test subject prior to the conduct of the screening test.

(a) Taste Threshold Screening

The Bitrex taste threshold screening, performed without wearing a respirator, is intended to determine whether the individual being tested can detect the taste of Bitrex.

(1) During threshold screening as well as during fit testing, subjects shall wear an enclosure about the head and shoulders that is approximately 12 inches (30.5 cm) in diameter by 14 inches (35.6 cm) tall. The front portion of the enclosure shall be clear from the respirator and allow free movement of the head when a respirator is worn. An enclosure substantially similar to the 3M hood assembly, parts # FT 14 and # FT 15 combined, is adequate.

(2) The test enclosure shall have a 3/4-inch (1.9-cm) hole in front of the test subject's nose and mouth area to accommodate the nebulizer nozzle.

(3) The test subject shall don the test enclosure. Throughout the threshold screening test, the test subject shall breathe through his or her slightly open mouth with tongue extended. The subject is instructed to report when he/she detects a bitter taste

(4) Using a DeVilbiss Model 40 Inhalation Medication Nebulizer or equivalent, the test conductor shall spray the Threshold Check Solution into the enclosure. This Nebulizer shall be clearly marked to distinguish it from the fit test solution nebulizer.

(5) The Threshold Check Solution is prepared by adding 13.5 milligrams of Bitrex to 100 ml of 5% salt (NaCl) solution in distilled water.

(6) To produce the aerosol, the nebulizer bulb is firmly squeezed so that the bulb collapses completely, and is then released and allowed to fully expand.

(7) An initial ten squeezes are repeated rapidly and then the test subject is asked whether the Bitrex can be tasted. If the test subject reports tasting the bitter taste during the ten squeezes, the screening test is completed. The taste threshold is noted as ten regardless of the number of squeezes actually completed.

(8) If the first response is negative, ten more squeezes are repeated rapidly and the test subject is again asked whether the Bitrex is tasted. If the test subject reports tasting the bitter taste during the second ten squeezes, the screening test is completed. The taste threshold is noted as twenty regardless of the number of squeezes actually completed.

(9) If the second response is negative, ten more squeezes are repeated rapidly and the test subject is again asked whether the Bitrex is tasted. If the test subject reports tasting the bitter taste during the third set of ten squeezes, the screening test is completed. The taste threshold is noted as thirty regardless of the number of squeezes actually completed.

(10) The test conductor will take note of the number of squeezes required to solicit a taste response.

(11) If the Bitrex is not tasted after 30 squeezes (step 10), the test subject is unable to taste Bitrex and may not perform the Bitrex fit test.

(12) If a taste response is elicited, the test subject shall be asked to take note of the taste for reference in the fit test.

(13) Correct use of the nebulizer means that approximately 1 ml of liquid is used at a time in the nebulizer body.

(14) The nebulizer shall be thoroughly rinsed in water, shaken to dry, and refilled at least each morning and afternoon or at least every four hours.

(b) Bitrex Solution Aerosol Fit Test Procedure

(1) The test subject may not eat, drink (except plain water), smoke, or chew gum for 15 minutes before the test.

(2) The fit test uses the same enclosure as that described in 4.(a) above.

(3) The test subject shall don the enclosure while wearing the respirator selected according to section I.A. of this appendix. The respirator shall be properly adjusted and equipped with any type particulate filter(s).

(4) A second DeVilbiss Model 40 Inhalation Medication Nebulizer or equivalent is used to spray the fit test solution into the enclosure. This nebulizer shall be clearly marked to distinguish it from the screening test solution nebulizer.

(5) The fit test solution is prepared by adding 337.5 mg of Bitrex to 200 ml of a 5% salt (NaCl) solution in warm water.

(6) As before, the test subject shall breathe through his or her slightly open mouth with tongue extended, and be instructed to report if he/she tastes the bitter taste of Bitrex.

(7) The nebulizer is inserted into the hole in the front of the enclosure and an initial concentration of the fit test solution is sprayed into the enclosure using the same number of squeezes (either 10, 20 or 30

squeezes) based on the number of squeezes required to elicit a taste response as noted during the screening test.

(8) After generating the aerosol, the test subject shall be instructed to perform the exercises in section I.A.14. of this appendix.

(9) Every 30 seconds the aerosol concentration shall be replenished using one half the number of squeezes used initially (e.g., 5, 10 or 15).

(10) The test subject shall indicate to the test conductor if at any time during the fit test the taste of Bitrex is detected. If the test subject does not report tasting the Bitrex, the test is passed.

(11) If the taste of Bitrex is detected, the fit is deemed unsatisfactory and the test is failed. A different respirator shall be tried and the entire test procedure is repeated (taste threshold screening and fit testing).

5. Irritant Smoke (Stannic Chloride) Protocol

This qualitative fit test uses a person's response to the irritating chemicals released in the "smoke" produced by a stannic chloride ventilation smoke tube to detect leakage into the respirator.

(a) General Requirements and Precautions

(1) The respirator to be tested shall be equipped with high efficiency particulate air (HEPA) or P100 series filter(s).

(2) Only stannic chloride smoke tubes shall be used for this protocol.

(3) No form of test enclosure or hood for the test subject shall be used.

(4) The smoke can be irritating to the eyes, lungs, and nasal passages. The test conductor shall take precautions to minimize the test subject's exposure to irritant smoke. Sensitivity varies, and certain individuals may respond to a greater degree to irritant smoke. Care shall be taken when performing the sensitivity screening checks that determine whether the test subject can detect irritant smoke to use only the minimum amount of smoke necessary to elicit a response from the test subject.

(5) The fit test shall be performed in an area with adequate ventilation to prevent exposure of the person conducting the fit test or the buildup of irritant smoke in the general atmosphere.

(b) Sensitivity Screening Check

The person to be tested must demonstrate his or her ability to detect a weak concentration of the irritant smoke.

(1) The test operator shall break both ends of a ventilation smoke tube containing stannic chloride, and attach one end of the smoke tube to a low flow air pump set to deliver 200 milliliters per minute, or an aspirator squeeze bulb. The test operator shall cover the other end of the smoke tube with a short piece of tubing to prevent potential injury from the jagged end of the smoke tube.

(2) The test operator shall advise the test subject that the smoke can be irritating to the eyes, lungs, and nasal passages and instruct the subject to keep his/her eyes closed while the test is performed.

(3) The test subject shall be allowed to smell a weak concentration of the irritant smoke before the respirator is donned to become familiar with its irritating properties and to determine if he/she can detect the irritating properties of the smoke. The test operator shall carefully direct a small amount of the irritant smoke in the test subject's direction to determine that he/she can detect it.

(c) Irritant Smoke Fit Test Procedure

(1) The person being fit tested shall don the respirator without assistance, and perform the required user seal check(s).

(2) The test subject shall be instructed to keep his/her eyes closed.

(3) The test operator shall direct the stream of irritant smoke from the smoke tube toward the faceseal area of the test subject, using the low flow pump or the squeeze bulb. The test operator shall begin at least 12 inches from the facepiece and move the smoke stream around the whole perimeter of the mask. The operator shall gradually make two more passes around the perimeter of the mask, moving to within six inches of the respirator.

(4) If the person being tested has not had an involuntary response and/or detected the irritant smoke, proceed with the test exercises.

(5) The exercises identified in section I.A.14. of this appendix shall be performed by the test subject while the respirator seal is being continually challenged by the smoke, directed around the perimeter of the respirator at a distance of six inches.

(6) If the person being fit tested reports detecting the irritant smoke at any time, the test is failed. The person being retested must repeat the entire sensitivity check and fit test procedure.

(7) Each test subject passing the irritant smoke test without evidence of a response (involuntary cough, irritation) shall be given a second sensitivity screening check, with the smoke from the same smoke tube

used during the fit test, once the respirator has been removed, to determine whether he/she still reacts to the smoke. Failure to evoke a response shall void the fit test.

(8) If a response is produced during this second sensitivity check, then the fit test is passed.

C. Quantitative Fit Test (QNFT) Protocols

The following quantitative fit testing procedures have been demonstrated to be acceptable: Quantitative fit testing using a non-hazardous test aerosol (such as corn oil, polyethylene glycol 400 [PEG 400], di-2-ethyl hexyl sebacate [DEHS], or sodium chloride) generated in a test chamber, and employing instrumentation to quantify the fit of the respirator; Quantitative fit testing using ambient aerosol as the test agent and appropriate instrumentation (condensation nuclei counter) to quantify the respirator fit; quantitative fit testing using controlled negative pressure and appropriate instrumentation to measure the volumetric leak rate of a facepiece to quantify the respirator fit.

1. General

(a) The employer shall ensure that persons administering QNFT are able to calibrate equipment and perform tests properly, recognize invalid tests, calculate fit factors properly and ensure that test equipment is in proper working order.

(b) The employer shall ensure that QNFT equipment is kept clean, and is maintained and calibrated according to the manufacturer's instructions so as to operate at the parameters for which it was designed.

2. Generated Aerosol Quantitative Fit Testing Protocol

(a) Apparatus

(1) Instrumentation. Aerosol generation, dilution, and measurement systems using particulates (corn oil, polyethylene glycol 400 [PEG 400], di-2-ethyl hexyl sebacate [DEHS] or sodium chloride) as test aerosols shall be used for quantitative fit testing.

(2) Test chamber. The test chamber shall be large enough to permit all test subjects to perform freely all required exercises without disturbing the test agent concentration or the measurement apparatus. The test chamber shall be equipped and constructed so that the test agent is effectively isolated from the ambient air, yet uniform in concentration throughout the chamber.

(3) When testing air-purifying respirators, the normal filter or cartridge element shall be replaced with a high efficiency particulate air (HEPA) or P100 series filter supplied by the same manufacturer.

(4) The sampling instrument shall be selected so that a computer record or strip chart record may be made of the test showing the rise and fall of the test agent concentration with each inspiration and expiration at fit factors of at least 2000. Integrators or computers that integrate the amount of test agent penetration leakage into the respirator for each exercise may be used provided a record of the readings is made.

(5) The combination of substitute air-purifying elements, test agent and test agent concentration shall be such that the test subject is not exposed in excess of an established exposure limit for the test agent at any time during the testing process, based upon the length of the exposure and the exposure limit duration.

(6) The sampling port on the test specimen respirator shall be placed and constructed so that no leakage occurs around the port (e.g., where the respirator is probed), a free air flow is allowed into the sampling line at all times, and there is no interference with the fit or performance of the respirator. The in-mask sampling device (probe) shall be designed and used so that the air sample is drawn from the breathing zone of the test subject, midway between the nose and mouth and with the probe extending into the facepiece cavity at least 1/2-inch.

(7) The test setup shall permit the person administering the test to observe the test subject inside the chamber during the test.

(8) The equipment generating the test atmosphere shall maintain the concentration of test agent constant to within a 10 percent variation for the duration of the test.

(9) The time lag (interval between an event and the recording of the event on the strip chart or computer or integrator) shall be kept to a minimum. There shall be a clear association between the occurrence of an event and its being recorded.

(10) The sampling line tubing for the test chamber atmosphere and for the respirator sampling port shall be of equal diameter and of the same material. The length of the two lines shall be equal.

(11) The exhaust flow from the test chamber shall pass through an appropriate filter (i.e., high efficiency particulate filter) before release.

(12) When sodium chloride aerosol is used, the relative humidity inside the test chamber shall not exceed 50 percent.

(13) The limitations of instrument detection shall be taken into account when determining the fit factor.

(14) Test respirators shall be maintained in proper working order and be inspected regularly for deficiencies such as cracks or missing valves and gaskets.

(b) Procedural Requirements

(1) When performing the initial user seal check using a positive or negative pressure check, the sampling line shall be crimped closed in order to avoid air pressure leakage during either of these pressure checks.

(2) The use of an abbreviated screening QLFT test is optional. Such a test may be utilized in order to quickly identify poor fitting respirators that passed the positive and/or negative pressure test and reduce the amount of QNFT time. The use of the CNC QNFT instrument in the count mode is another optional method to obtain a quick estimate of fit and eliminate poor fitting respirators before going on to perform a full QNFT.

(3) A reasonably stable test agent concentration shall be measured in the test chamber prior to testing. For canopy or shower curtain types of test units, the determination of the test agent's stability may be established after the test subject has entered the test environment.

(4) Immediately after the subject enters the test chamber, the test agent concentration inside the respirator shall be measured to ensure that the peak penetration does not exceed 5 percent for a half mask or 1 percent for a full facepiece respirator.

(5) A stable test agent concentration shall be obtained prior to the actual start of testing.

(6) Respirator restraining straps shall not be over-tightened for testing. The straps shall be adjusted by the wearer without assistance from other persons to give a reasonably comfortable fit typical of normal use. The respirator shall not be adjusted once the fit test exercises begin.

(7) The test shall be terminated whenever any single peak penetration exceeds 5 percent for half masks and 1 percent for full facepiece respirators. The test subject shall be refitted and retested.

(8) Calculation of fit factors

(i) The fit factor shall be determined for the quantitative fit test by taking the ratio of the average chamber concentration to the concentration measured inside the respirator for each test exercise except the grimace exercise.

(ii) The average test chamber concentration shall be calculated as the arithmetic average of the concentration measured before and after each test (i.e., 7 exercises) or the arithmetic average of the concentration measured before and after each exercise or the true average measured continuously during the respirator sample.

(iii) The concentration of the challenge agent inside the respirator shall be determined by one of the following methods:

(A) Average peak penetration method means the method of determining test agent penetration into the respirator utilizing a strip chart recorder, integrator, or computer. The agent penetration is determined by an average of the peak heights on the graph or by computer integration, for each exercise except the grimace exercise. Integrators or computers that calculate the actual test agent penetration into the respirator for each exercise will also be considered to meet the requirements of the average peak penetration method.

(B) Maximum peak penetration method means the method of determining test agent penetration in the respirator as determined by strip chart recordings of the test. The highest peak penetration for a given exercise is taken to be representative of average penetration into the respirator for that exercise.

(C) Integration by calculation of the area under the individual peak for each exercise except the grimace exercise. This includes computerized integration.

(D) The calculation of the overall fit factor using individual exercise fit factors involves first converting the exercise fit factors to penetration values, determining the average, and then converting that result back to a fit factor. This procedure is described in the following equation:

$$\text{overall fit factor} = \frac{\text{number of exercises}}{1/ff_1 + 1/ff_2 + 1/ff_3 + 1/ff_4 + 1/ff_5 + 1/ff_7 + 1/ff_8}$$

where ff_1, ff_2, ff_3, etc. are the fit factors for exercises 1, 2, 3, etc.

(9) The test subject shall not be permitted to wear a half mask or quarter facepiece respirator unless a minimum fit factor of 100 is obtained, or a full facepiece respirator unless a minimum fit factor of 500 is obtained.

(10) Filters used for quantitative fit testing shall be replaced whenever increased breathing resistance is encountered, or when the test agent has altered the integrity of the filter media.

3. Ambient aerosol condensation nuclei counter (CNC) quantitative fit testing protocol.

The ambient aerosol condensation nuclei counter (CNC) quantitative fit testing (Portacount™) protocol quantitatively fit tests respirators with the use of a probe. The probed respirator is only used for quantitative fit tests. A probed respirator has a special sampling device, installed on the respirator, that allows the probe to sample the air from inside the mask. A probed respirator is required for each make, style, model, and size that the employer uses and can be obtained from the respirator manufacturer or distributor. The CNC instrument manufacturer, TSI Inc., also provides probe attachments (TSI sampling adapters) that permit fit testing in an employee's own respirator. A minimum fit factor pass level of at least 100 is necessary for a half-mask respirator and a minimum fit factor pass level of at least 500 is required for a full facepiece negative pressure respirator. The entire screening and testing procedure shall be explained to the test subject prior to the conduct of the screening test.

(a) Portacount Fit Test Requirements

(1) Check the respirator to make sure the sampling probe and line are properly attached to the facepiece and that the respirator is fitted with a particulate filter capable of preventing significant penetration by the ambient particles used for the fit test (e.g., NIOSH 42 CFR 84 series 100, series 99, or series 95 particulate filter) per manufacturer's instruction.

(2) Instruct the person to be tested to don the respirator for five minutes before the fit test starts. This purges the ambient particles trapped inside the respirator and permits the wearer to make certain the respirator is comfortable. This individual shall already have been trained on how to wear the respirator properly.

(3) Check the following conditions for the adequacy of the respirator fit: Chin properly placed; Adequate strap tension, not overly tightened; Fit across nose bridge; Respirator of proper size to span distance from nose to chin; Tendency of the respirator to slip; Self-observation in a mirror to evaluate fit and respirator position.

(4) Have the person wearing the respirator do a user seal check. If leakage is detected, determine the cause. If leakage is from a poorly fitting facepiece, try another size of the same model respirator, or another model of respirator.

(5) Follow the manufacturer's instructions for operating the Portacount and proceed with the test.

(6) The test subject shall be instructed to perform the exercises in section I.A.14. of this appendix.

(7) After the test exercises, the test subject shall be questioned by the test conductor regarding the comfort of the respirator upon completion of the protocol. If it has become unacceptable, another model of respirator shall be tried.

(b) Portacount Test Instrument

(1) The Portacount will automatically stop and calculate the overall fit factor for the entire set of exercises. The overall fit factor is what counts. The Pass or Fail message will indicate whether or not the test was successful. If the test was a Pass, the fit test is over.

(2) Since the pass or fail criterion of the Portacount is user programmable, the test operator shall ensure that the pass or fail criterion meet the requirements for minimum respirator performance in this Appendix.

(3) A record of the test needs to be kept on file, assuming the fit test was successful. The record must contain the test subject's name; overall fit factor; make, model, style, and size of respirator used; and date tested.

4. Controlled negative pressure (CNP) quantitative fit testing protocol.

The CNP protocol provides an alternative to aerosol fit test methods. The CNP fit test method technology is based on exhausting air from a temporarily sealed respirator facepiece to generate and then maintain a constant negative pressure inside the facepiece. The rate of air exhaust is controlled so that a constant negative pressure is maintained in the respirator during the fit test. The level of pressure is selected to replicate the mean inspiratory pressure that causes leakage into the respirator under normal use conditions. With pressure held constant, air flow out of the respirator is equal to air flow into the respirator. Therefore, measurement of the exhaust stream that is required to hold the pressure in the temporarily sealed respirator constant yields a direct measure of leakage air flow into the respirator. The CNP fit test method measures leak rates through the facepiece as a method for determining the facepiece fit for negative pressure respirators. The CNP instrument manufacturer Dynatech Nevada also provides attachments (sampling manifolds) that replace the filter cartridges to permit fit testing in an employee's own respirator. To perform the test, the test subject closes his or her mouth and holds his/her breath, after which an air pump removes air from the respirator facepiece at a pre-selected constant pressure. The facepiece fit is expressed as the leak rate through the facepiece, expressed as milliliters per minute. The quality and validity of the CNP fit tests are determined by the degree to which the in-mask pressure tracks the test pressure during the system measurement time of approximately five seconds. Instantaneous feedback in the form of a real-time pressure trace of the in-mask pressure is provided and used to determine test validity and quality. A minimum fit factor pass level of 100 is necessary for a half-mask respirator and a minimum fit factor of at least 500 is required for a full facepiece respirator. The entire screening and testing procedure shall be explained to the test subject prior to the conduct of the screening test.

(a) CNP Fit Test Requirements.

(1) The instrument shall have a non-adjustable test pressure of 15.0 mm water pressure.

(2) The CNP system defaults selected for test pressure shall be set at 15 mm of water (0.58 inches of water) and the modeled inspiratory flow rate shall be 53.8 liters per minute for performing fit tests.

Note: CNP systems have built-in capability to conduct fit testing that is specific to unique work rate, mask, and gender situations that might apply in a specific workplace. Use of system default values, which were selected to represent respirator wear with medium cartridge resistance at a low-moderate work rate, will allow inter-test comparison of the respirator fit.

(3) The individual who conducts the CNP fit testing shall be thoroughly trained to perform the test.

(4) The respirator filter or cartridge needs to be replaced with the CNP test manifold. The inhalation valve downstream from the manifold either needs to be temporarily removed or propped open.

(5) The test subject shall be trained to hold his or her breath for at least 20 seconds.

(6) The test subject shall don the test respirator without any assistance from the individual who conducts the CNP fit test.

(7) The QNFT protocol shall be followed according to section I.C.1 of this appendix with an exception for the CNP test exercises.

(b) CNP Test Exercises

(1) Normal breathing. In a normal standing position, without talking, the subject shall breathe normally for 1 minute. After the normal breathing exercise, the subject needs to hold head straight ahead and hold his or her breath for 10 seconds during the test measurement.

(2) Deep breathing. In a normal standing position, the subject shall breathe slowly and deeply for 1 minute, being careful not to hyperventilate. After the deep breathing exercise, the subject shall hold his or her head straight ahead and hold his or her breath for 10 seconds during test measurement.

(3) Turning head side to side. Standing in place, the subject shall slowly turn his or her head from side to side between the extreme positions on each side for 1 minute. The head shall be held at each extreme momentarily so the subject can inhale at each side. After the turning head side to side exercise, the subject needs to hold head full left and hold his or her breath for 10 seconds during test measurement. Next, the subject needs to hold head full right and hold his or her breath for 10 seconds during test measurement.

(4) Moving head up and down. Standing in place, the subject shall slowly move his or her head up and down for 1 minute. The subject shall be instructed to inhale in the up position (i.e., when looking toward the ceiling). After the moving head up and down exercise, the subject shall hold his or her head full up and hold his or her breath for 10 seconds during test measurement. Next, the subject shall hold his or her head full down and hold his or her breath for 10 seconds during test measurement.

(5) Talking. The subject shall talk out loud slowly and loud enough so as to be heard clearly by the test conductor. The subject can read from a prepared text such as the Rainbow Passage, count backward from 100, or recite a memorized poem or song for 1 minute. After the talking exercise, the subject shall hold his or her head straight ahead and hold his or her breath for 10 seconds during the test measurement.

(6) Grimace. The test subject shall grimace by smiling or frowning for 15 seconds.

(7) Bending over. The test subject shall bend at the waist as if he or she were to touch his or her toes for 1 minute. Jogging in place shall be substituted for this exercise in those test environments such as shroud-type QNFT units that prohibit bending at the waist. After the bending over exercise, the subject shall hold his or her head straight ahead and hold his or her breath for 10 seconds during the test measurement.

(8) Normal Breathing. The test subject shall remove and re-don the respirator within a one-minute period. Then, in a normal standing position, without talking, the subject shall breathe normally for 1 minute. After the normal breathing exercise, the subject shall hold his or her head straight ahead and hold his or her breath for 10 seconds during

the test measurement. After the test exercises, the test subject shall be questioned by the test conductor regarding the comfort of the respirator upon completion of the protocol. If it has become unacceptable, another model of a respirator shall be tried

(c) CNP Test Instrument

(1) The test instrument shall have an effective audio warning device when the test subject fails to hold his or her breath during the test. The test shall be terminated whenever the test subject failed to hold his or her breath. The test subject may be refitted and retested.

(2) A record of the test shall be kept on file, assuming the fit test was successful. The record must contain the test subject's name; overall fit factor; make, model, style and size of respirator used; and date tested.

Part II. New Fit Test Protocols

A. Any person may submit to OSHA an application for approval of a new fit test protocol. If the application meets the following criteria, OSHA will initiate a rulemaking proceeding under section 6(b)(7) of the OSH Act to determine whether to list the new protocol as an approved protocol in this Appendix A.

B. The application must include a detailed description of the proposed new fit test protocol. This application must be supported by either:

1. A test report prepared by an independent government research laboratory (e.g., Lawrence Livermore National Laboratory, Los Alamos National Laboratory, the National Institute for Standards and Technology) stating that the laboratory has tested the protocol and had found it to be accurate and reliable; or

2. An article that has been published in a peer-reviewed industrial hygiene journal describing the protocol and explaining how test data support the protocol's accuracy and reliability.

C. If OSHA determines that additional information is required before the Agency commences a rulemaking proceeding under this section, OSHA will so notify the applicant and afford the applicant the opportunity to submit the supplemental information. Initiation of a rulemaking proceeding will be deferred until OSHA has received and evaluated the supplemental information.

[63 FR 20098, April 23, 1998]

Appendix 1B1

Appendix B1 to § 1910.134 (Mandatory): User Seal Check Procedures

The individual who uses a tight-fitting respirator is to perform a user seal check to ensure that an adequate seal is achieved each time the respirator is put on. Either the positive and negative pressure checks listed in this appendix, or the respirator manufacturer's recommended user seal check method shall be used. User seal checks are not substitutes for qualitative or quantitative fit tests.

I. Facepiece Positive and/or Negative Pressure Checks

A. *Positive pressure check.* Close off the exhalation valve and exhale gently into the facepiece. The face fit is considered satisfactory if a slight positive pressure can be built up inside the facepiece without any evidence of outward leakage of air at the seal. For most respirators this method of leak testing requires the wearer to first remove the exhalation valve cover before closing off the exhalation valve and then carefully replacing it after the test.

B. *Negative pressure check.* Close off the inlet opening of the canister or cartridge(s) by covering with the palm of the hand(s) or by replacing the filter seal(s), inhale gently so that the facepiece collapses slightly, and hold the breath for ten seconds. The design of the inlet opening of some cartridges cannot be effectively covered with the palm of the hand. The test can be performed by covering the inlet opening of the cartridge with a thin latex or nitrile glove. If the facepiece remains in its slightly collapsed condition and no inward leakage of air is detected, the tightness of the respirator is considered satisfactory.

II. Manufacturer's Recommended User Seal Check Procedures

The respirator manufacturer's recommended procedures for performing a user seal check may be used instead of the positive and/or negative pressure check procedures provided that the employer demonstrates that the manufacturer's procedures are equally effective.

[63 FR 1152, Jan. 8, 1998]

Appendix 1B2

Appendix B2 to § 1910.134 (Mandatory): Respirator Cleaning Procedures

These procedures are provided for employer use when cleaning respirators. They are general in nature, and the employer as an alternative may use the cleaning recommendations provided by the manufacturer of the respirators used by their employees, provided such procedures are as effective as those listed here in Appendix B2. Equivalent effectiveness simply means that the procedures used must accomplish the objectives set forth in Appendix B2, i.e., must ensure that the respirator is properly cleaned and disinfected in a manner that prevents damage to the respirator and does not cause harm to the user.

I. Procedures for Cleaning Respirators

A. Remove filters, cartridges, or canisters. Disassemble facepieces by removing speaking diaphragms, demand and pressure- demand valve assemblies, hoses, or any components recommended by the manufacturer. Discard or repair any defective parts.

B. Wash components in warm (43°C [110°F] maximum) water with a mild detergent or with a cleaner recommended by the manufacturer. A stiff bristle (not wire) brush may be used to facilitate the removal of dirt.

C. Rinse components thoroughly in clean, warm (43°C [110°F] maximum), preferably running water. Drain.

D. When the cleaner used does not contain a disinfecting agent, respirator components should be immersed for two minutes in one of the following:

1. Hypochlorite solution (50 ppm of chlorine) made by adding approximately one milliliter of laundry bleach to one liter of water at 43°C (110°F); or,

2. Aqueous solution of iodine (50 ppm iodine) made by adding approximately 0.8 milliliters of tincture of iodine (6-8 grams ammonium and/or potassium iodide/100 cc of 45% alcohol) to one liter of water at 43°C (110°F); or,

3. Other commercially available cleansers of equivalent disinfectant quality when used as directed, if their use is recommended or approved by the respirator manufacturer.

E. Rinse components thoroughly in clean, warm (43°C [110°F] maximum), preferably running water. Drain. The importance of thorough rinsing cannot be overemphasized. Detergents or disinfectants that dry on facepieces may result in dermatitis. In addition, some disinfectants may cause deterioration of rubber or corrosion of metal parts if not completely removed.

F. Components should be hand-dried with a clean lint-free cloth or air-dried.

G. Reassemble facepiece, replacing filters, cartridges, and canisters where necessary.

H. Test the respirator to ensure that all components work properly.

[63 FR 1152, Jan. 8, 1998]

Appendix 1C

Appendix C to § 1910.134 (Mandatory): OSHA Respirator Medical Evaluation Questionnaire (English)

To the employer: Answers to questions in Section 1, and to question 9 in Section 2 of Part A, do not require a medical examination.

To the employee:

Can you read (circle one): Yes/No

Your employer must allow you to answer this questionnaire during normal working hours, or at a time and place that is convenient to you. To maintain your confidentiality, your employer or supervisor must not look at or review your answers, and your employer must tell you how to deliver or send this questionnaire to the health care professional who will review it.

Part A. Section 1 (Mandatory). The following information must be provided by every employee who has been selected to use any type of respirator (please print).

1. Today's date: ______________________________

2. Your name: ______________________________

3. Your age (to nearest year: ______________________________

4. Sex (circle one): Male/Female

5. Your height: __________ ft. in. __________

6. Your weight: __________ lbs.

7. Your job title: ______________________________

8. A phone number where you can be reached by the health care professional who reviews this questionnaire (include the Area Code):

   ____

9. The best time to phone you at this number: ______________

10. Has your employer told you how to contact the health care professional who will review this questionnaire (circle one): Yes/No

11. Check the type of respirator you will use (you can check more than one category):

    a. ________ N, R, or P disposable respirator (filter-mask, non-cartridge type only).

    b. ________ Other type (for example, half- or full-facepiece type, powered-air purifying, supplied-air, self-contained breathing apparatus).

12. Have you worn a respirator (circle one): Yes/No

    If "yes," what type(s) ______________________________

    ____

Part A. Section 2 (Mandatory). Questions 1 through 9 below must be answered by every employee who has been selected to use any type of respirator (please circle "yes" or "no").

1. Do you *currently* smoke tobacco, or have you smoked tobacco in the last month: Yes/No

2. Have you *ever had* any of the following conditions?

   a. Seizures (fits): Yes/No
   b. Diabetes (sugar disease): Yes/No
   c. Allergic reactions that interfere with your breathing: Yes/No
   d. Claustrophobia (fear of closed-in places): Yes/No
   e. Trouble smelling odors: Yes/No

3. Have you *ever had* any of the following pulmonary or lung problems?

   a. Asbestosis: Yes/No
   b. Asthma: Yes/No

c. Chronic bronchitis: Yes/No
d. Emphysema: Yes/No
e. Pneumonia: Yes/No
f. Tuberculosis: Yes/No
g. Silicosis: Yes/No
h. Pneumothorax (collapsed lung): Yes/No
i. Lung cancer: Yes/No
j. Broken ribs: Yes/No
k. Any chest injuries or surgeries: Yes/No
l. Any other lung problem that you've been told about: Yes/No

4. Do you *currently* have any of the following symptoms of pulmonary or lung illness?

a. Shortness of breath: Yes/No
b. Shortness of breath when walking fast on level ground or walking up a slight hill or incline: Yes/No
c. Shortness of breath when walking with other people at an ordinary pace on level ground: Yes/No
d. Have to stop for breath when walking at your own pace on level ground: Yes/No
e. Shortness of breath when washing or dressing yourself: Yes/No
f. Shortness of breath that interferes with your job: Yes/No
g. Coughing that produces phlegm (thick sputum): Yes/No
h. Coughing that wakes you early in the morning: Yes/No
i. Coughing that occurs mostly when you are lying down: Yes/No
j. Coughing up blood in the last month: Yes/No
k. Wheezing: Yes/No
l. Wheezing that interferes with your job: Yes/No
m. Chest pain when you breathe deeply: Yes/No
n. Any other symptoms that you think may be related to lung problems: Yes/No

5. Have you *ever had* any of the following cardiovascular or heart problems?

a. Heart attack: Yes/No
b. Stroke: Yes/No
c. Angina: Yes/No
d. Heart failure: Yes/No
e. Swelling in your legs or feet (not caused by walking): Yes/No
f. Heart arrhythmia (heart beating irregularly): Yes/No
g. High blood pressure: Yes/No
h. Any other heart problem that you've been told about: Yes/No

6. Have you *ever had* any of the following cardiovascular or heart symptoms?

   a. Frequent pain or tightness in your chest: Yes/No
   b. Pain or tightness in your chest during physical activity: Yes/No
   c. Pain or tightness in your chest that interferes with your job: Yes/No
   d. In the past two years, have you noticed your heart skipping or missing a beat: Yes/No
   e. Heartburn or indigestion that is not related to eating: Yes/ No
   f. Any other symptoms that you think may be related to heart or circulation problems: Yes/No

7. Do you *currently* take medication for any of the following problems?

   a. Breathing or lung problems: Yes/No
   b. Heart trouble: Yes/No
   c. Blood pressure: Yes/No
   d. Seizures (fits): Yes/No

8. If you've used a respirator, have you *ever had* any of the following problems? (If you've never used a respirator, check the following space and go to question 9:)

   a. Eye irritation: Yes/No
   b. Skin allergies or rashes: Yes/No
   c. Anxiety: Yes/No
   d. General weakness or fatigue: Yes/No
   e. Any other problem that interferes with your use of a respirator: Yes/No

9. Would you like to talk to the health care professional who will review this questionnaire about your answers to this questionnaire: Yes/No

Questions 10 to 15 below must be answered by every employee who has been selected to use either a full-facepiece respirator or a self-contained breathing apparatus (SCBA). For employees who have been selected to use other types of respirators, answering these questions is voluntary.

10. Have you ever lost vision in either eye (temporarily or permanently): Yes/No

11. Do you currently have any of the following vision problems?
    a. Wear contact lenses: Yes/No
    b. Wear glasses: Yes/No
    c. Color blind: Yes/No
    d. Any other eye or vision problem: Yes/No

12. Have you ever had an injury to your ears, including a broken ear drum: Yes/No

13. Do you currently have any of the following hearing problems?
    a. Difficulty hearing: Yes/No
    b. Wear a hearing aid: Yes/No
    c. Any other hearing or ear problem: Yes/No

14. Have you ever had a back injury: Yes/No

15. Do you currently have any of the following musculoskeletal problems?
    a. Weakness in any of your arms, hands, legs, or feet: Yes/No
    b. Back pain: Yes/No
    c. Difficulty fully moving your arms and legs: Yes/No
    d. Pain or stiffness when you lean forward or backward at the waist: Yes/No
    e. Difficulty fully moving your head up or down: Yes/No
    f. Difficulty fully moving your head side to side: Yes/No
    g. Difficulty bending at your knees: Yes/No
    h. Difficulty squatting to the ground: Yes/No
    i. Climbing a flight of stairs or a ladder carrying more than 25 lbs: Yes/No
    j. Any other muscle or skeletal problem that interferes with using a respirator: Yes/No

Part B. Any of the following questions, and other questions not listed, may be added to the questionnaire at the discretion of the health care professional who will review the questionnaire.

1. In your present job, are you working at high altitudes (over 5,000 feet) or in a place that has lower than normal amounts of oxygen: Yes/No

If "yes," do you have feelings of dizziness, shortness of breath, pounding in your chest, or other symptoms when you're working under these conditions: Yes/No ______________________________

____

____

2. At work or at home, have you ever been exposed to hazardous solvents, hazardous airborne chemicals (e.g., gases, fumes, or dust), or have you come into skin contact with hazardous chemicals: Yes/No

   If "yes," name the chemicals if you know them: ______________

   ____

   ____

3. Have you ever worked with any of the materials, or under any of the conditions, listed below:

   a. Asbestos: Yes/No
   b. Silica (e.g., in sandblasting): Yes/No
   c. Tungsten/cobalt (e.g., grinding or welding this material): Yes/No
   d. Beryllium: Yes/No
   e. Aluminum: Yes/No
   f. Coal (for example, mining): Yes/No
   g. Iron: Yes/No
   h. Tin: Yes/No
   i. Dusty environments: Yes/No
   j. Any other hazardous exposures: Yes/No

   If "yes," describe these exposures: ______________________

   ____

   ____

4. List any second jobs or side businesses you have: ______________

   ____

   ____

5. List your previous occupations: ______________________

   ____

6. List your current and previous hobbies: ____________________

____

7. Have you been in the military services? Yes/No

   If "yes," were you exposed to biological or chemical agents (either in training or combat): Yes/No

8. Have you ever worked on a HAZMAT team? Yes/No

9. Other than medications for breathing and lung problems, heart trouble, blood pressure, and seizures mentioned earlier in this questionnaire, are you taking any other medications for any reason (including over-the-counter medications): Yes/No

   If "yes," name the medications if you know them: ____________

10. Will you be using any of the following items with your respirator(s)?

    a. HEPA Filters: Yes/No
    b. Canisters (for example, gas masks): Yes/No
    c. Cartridges: Yes/No

11. How often are you expected to use the respirator(s) (circle "yes" or "no" for all answers that apply to you)?:

    a. Escape only (no rescue): Yes/No
    b. Emergency rescue only: Yes/No
    c. Less than 5 hours per week: Yes/No
    d. Less than 2 hours per day: Yes/No
    e. 2 to 4 hours per day: Yes/No
    f. Over 4 hours per day: Yes/No

12. During the period you are using the respirator(s), is your work effort:

    a. Light (less than 200 kcal per hour): Yes/No

    If "yes," how long does this period last during the average shift:

    __________ hrs. __________ mins.

    Examples of a light work effort are sitting while writing, typing, drafting, or performing light assembly work; or standing while operating a drill press (1-3 lbs.) or controlling machines.

b. Moderate (200 to 350 kcal per hour): Yes/No

If "yes," how long does this period last during the average shift:

_________ hrs. _________ mins.

Examples of moderate work effort are sitting while nailing or filing; driving a truck or bus in urban traffic; standing while drilling, nailing, performing assembly work, or transferring a moderate load (about 35 lbs.) at trunk level; walking on a level surface about 2 mph or down a 5-degree grade about 3 mph; or pushing a wheelbarrow with a heavy load (about 100 lbs.) on a level surface.

c. Heavy (above 350 kcal per hour): Yes/No

If "yes," how long does this period last during the average shift:

_________ hrs. _________ mins.

Examples of heavy work are lifting a heavy load (about 50 lbs.) from the floor to your waist or shoulder; working on a loading dock; shoveling; standing while bricklaying or chipping castings; walking up an 8-degree grade about 2 mph; climbing stairs with a heavy load (about 50 lbs.).

13. Will you be wearing protective clothing and/or equipment (other than the respirator) when you're using your respirator: Yes/No

    If "yes," describe this protective clothing and/or equipment:

    ____

    ____

14. Will you be working under hot conditions (temperature exceeding 77°F): Yes/No

15. Will you be working under humid conditions: Yes/No

16. Describe the work you'll be doing while you're using your respirator(s):

    ____

    ____

17. Describe any special or hazardous conditions you might encounter when you're using your respirator(s) (for example, confined spaces, life-threatening gases):

____

____

18. Provide the following information, if you know it, for each toxic substance that you'll be exposed to when you're using your respirator(s):

Name of the first toxic substance: ____________________

Estimated maximum exposure level per shift: ______________

Duration of exposure per shift: ______________________

Name of the second toxic substance: __________________

Estimated maximum exposure level per shift: ______________

Duration of exposure per shift: ______________________

Name of the third toxic substance: ____________________

Estimated maximum exposure level per shift: ______________

Duration of exposure per shift: ______________________

The name of any other toxic substances that you'll be exposed to while using your respirator:

____

____

____

19. Describe any special responsibilities you'll have while using your respirator(s) that may affect the safety and well-being of others (for example, rescue, security):

____

____

Apéndice 1C

Cuestionario de Evaluación Médico obligado por la OSHA (La agencia de seguridad y salud ocupacional) Parte 29 CFR 1910.134 Mandatorio para Proteccion del Sistema Respiratorio (Español)

Marque con un circulo para indicar sus respuestas a cada pregunta.

Para el empleado: Puede usted leer (circule uno): Sí o No

Su patrón debe dejarlo responder estas preguntas durante horas de trabajo o en un tiempo y lugar que sea conveniente para usted. Para mantener este cuestionario confidencial, su patrón o supervisor no debe ver o revisar sus respuestas. Su patrón debe informarle a quien dar o enviar este cuestionario para ser revisado por un profesional de sanidad con licencia autorizado por el estado.

Parte A. Sección 1. (Mandatorio). La siguiente información debe de ser proveida por cada empleado que hasido seleccionado para usar cualquier tipo de respirador (escriba claro por favor).

1. Fecha : ______________________________
2. Nombre: ______________________________
3. Edad: ______________________________
4. Su sexo (circule uno): Masculino o Femenino
5. Altura: __________ pies __________ pulgadas
6. Peso: __________ libras
7. Su ocupación, título o tipo de trabajo: ______________________

8. Número de teléfono al donde pueda ser llamado por un profesional de sanidad con licencia que: ______________________

   revisara este cuestionario (incluya el área): ______________________

9. Indique la hora mas conveniente para llamarle a este numero:

   ____

10. ¿Le ha informado su patrón como comunicarse con el profesional de sanidad con licencia que va a revisareste cuestionario (circule una respuesta)? Sí o No

11. Anote el tipo de equipo protector respíratorio que va utilizar (puede anotar mas de una categoría)

    a. __________ Respirador disponible de clase N, R, o P (por ejemplo: respirador de filtro mécanico, respirador sin cartucho)
    b. __________ Otros tipos (respirador con cartucho químico, máscara con cartucho químico, máscara con manguera con soplador (PAPR),máscara con manguera sin soplador (SAR), aparato respiratorio autónomos (SCBA)).

12. ¿Ha usado algun tipo de respirador ? Sí o No

    Si ha usado equipo protector respíratorio, que tipo(s) ha utilizado:

    ____

    ____

Parte A. Seccion 2. (Mandatorio): Preguntas del 1 al 9 deben ser contestadas por cada empleado que fue seleccionado a usar cualquier tipo de respirador. Marque con un circulo para indicar sus repuestas.

1. ¿Corrientemente fuma tabaco, o ha fumado tabaco durante el ultimo mes? Sí o No

2. ¿Ha tenido algunas de las siguientes condiciones medicas?

   a. Convulsiones: Sí o No
   b. Diabetes (azucar en la sangre): Sí o No
   c. Reacciones alergicas que no lo deja respirar: Sí o No
   d. Claustrofobia (miedo de estar en espacios cerrados): Sí o No
   e. Dificultad oliendo excepto cuando ha cogido un resfriado: Sí o No

3. ¿Ha tenido algunas de los siguientes problemas pulmonares?
   a. Asbestosis: Sí o No
   b. Asma: Sí o No
   c. Bronquitis cronica: Sí o No
   d. Emfisema: Sí o No
   e. Pulmonía: Sí o No
   f. Tuberculosis: Sí o No
   g. Silicosis: Sí o No
   h. Neumotorax (pulmon colapsado): Sí o No
   i. Cáncer en los pulmones: Sí o No
   j. Costillas quebradas: Sí o No
   k. Injuria o cirujía en el pecho: Sí o No
   l. Algun otro problema de los pulmones que le ha dicho su medico: Sí o No

4. ¿Corrientemente tiene alguno de los siguientes síntomas o enfermedades en sus pulmones?
   a. Respiración dificultosa: Si o No
   b. Respiración dificultosa cuando camina rapido sobre terreno plano o subiendo una colina: Sí o No
   c. Respiración dificultosa cuando camina normalmente con otras personas sobre terreno plano: Sí o No
   d. Cuando camina normalmente en terreno plano se encuentra corto de resuello? Sí o No
   e. Respiración dificultosa cuando se esta bañando o vistiendo: Sí o No
   f. Respiración dificultosa que lo impede trabajar: Sí o No
   g. Tos con flema: Sí o No
   h. Tos que lo despierta temprano en la mañana: Sí o No
   i. Tos que occure cuando esta acostado: Sí o No
   j. Ha tosido sangre en el ultimo mes: Sí o No
   k. Silbar o respirar con mucha dificultad: Sí o No
   l. Silbar que lo impede trabajar: Sí o No
   m. Dolor del pecho cuando respira profundamente: Sí o No
   n. Otros símtomas que crea usted estar relacionados a los pulmones: Sí o No

5. ¿Ha tenido algunos de los siguientes problemas con el corazón?
   a. Ataque cardiaco: Sí o No
   b. Ataque cerebrovascular: Sí o No
   c. Dolor en el pecho: Sí o No
   d. Falla de corazón: Sí o No
   e. Hinchazón en las piernas o pies (que no sea por caminar): Sí o No

f. Latidos irregulares del corazón: Sí o No
g. Alta presión: Sí o No
h. Algun otro problema cardio-vascular o cardiaco: Sí o No

6. ¿Ha tenido algunos de los siguientes síntomas causados por su corazón?

a. Dolor de pecho frecuente o pecho apretado: Sí o No
b. Dolor o pecho apretado durante actividad fisica: Sí o No
c. Dolor o pecho apretado que no lo deja trabajar normalmente: Sí o No
d. En los ultimos dos años ha notado que su corazón late irregularmente: Sí o No
e. Dolor en el pecho o indigestion que no es relacionado a la comida: Sí o No
f. Algunos otros síntomas que usted piensa ser causado por problemas de su corazón o de su circulation: Sí o No

7. ¿Esta tomando medicina por algunso de los siguientes problemas?

a. Respiración dificultosa: Sí o No
b. Problemas del corazón: Sí o No
c. Alta presión: Sí o No
d. Convulsiones: Sí o No

8. ¿Le ha causado alguno de los siguientes problemas usando el respirador? (si no ha usado un respirador, dejeesta pregunta en blanco__________ y continue con pregunta

a. Irritación de los ojos: Sí o No
b. Alergias del cutis o sarpullido: Sí o No
c. Ansiedad que ocurre solamente cuando usa el respirado: Sí o No
d. Debilidad, falta de vigor o fatiga desacostumbrada: Sí o No
e. Algun otro problema que le impida utilizar su respirador: Sí o No

9. ¿Le gustaria hablar con el profesional de sanidad con licencia autorizado por el estado que revisara estecuestionario sobre sus respuestas? Sí o No

Las preguntas del 10 al 15 deben ser contestadas por los empleados seleccionados para usar una máscara con cartucho químico o aparato respiratorio autónomo (SCBA). Los empleados que usan otro tipo de respirador no tienen que contestar estas preguntas.

10. ¿Ha perdido la vista en cualquiera de sus ojos (temporalmente o permanente): Sí o No

11. ¿Corrientemente tiene algunos de los siguientes problemas con su vista?

    a. Usa lentes de contacto: Sí o No
    b. Usa lentes: Sí o No
    c. Daltoniano (dificultad distinguiendo colores): Sí o No
    d. Tiene algún problema con sus ojos o su vista: Sí o No

12. ¿Ha tenido daño en sus oidos incluyendo rotura del tímpano: Sí o No

13. ¿Corrientemente tiene uno de las siguientes problemas para oir?

    a. Dificultad oyendo: Sí o No
    b. Usa un aparato para oir: Sí o No
    c. Tiene algun otro problema con sus oidos o dificultad escuchando: Sí o No

14. ¿Se ha dañado o lastimado su espalda? Sí o No

15. ¿Tiene uno de los siguientes problemas de su aparato muscular or eskeleto?

    a. Debilidad en sus brazos, manos, piernas o pies: Sí o No
    b. Dolor de espalda: Sí o No
    c. Dificultad para mover sus brazos y piernas completamente: Sí o No
    d. Dolor o engarrotamiento cuando se inclina para adelante o para atras: Sí o No
    e. Dificultad para mover su cabeza para arriba o para abajo completamente: Sí o No
    f. Dificultad para mover su cabeza de lado a lado: Sí o No
    g. Dificultad para agacharse doblando sus rodillas: Sí o No
    h. Dificultad para agacharse hasta tocar el piso: Sí o No
    i. Dificultad subiendo escaleras cargando mas de 25 libras: Sí o No
    j. Alguno problema muscular o con sus huesos que le evite usar un respirador: Sí o No

Parte B . Las siguientes preguntas pueden ser agregadas al cuestionario a discrecion del profesional de sanidad con licencia autorizado por el estado.

1. ¿Esta trabajando en las alturas arriba de 5,000 pies o en sitios que tienen menos oxígeno de lo normal? Sí o No

Si la respuesta es "Sí", se ha sentido mareado, o ha tenido dificultad respirando, palpitaciones, o cualquier otro síntoma que usted no tiene cuando no esta trabajando bajo estas condiciones: Sí o No

____

____

____

2. ¿En el trabajo o en su casa, ha estado expuesto a solventes o contaminantes peligrosos en el aire (por ejemplo,humos, neblina o polvos) o ha tenido contacto del cutis con químicas peligrosas? Sí o No

   Escriba las químicas y productos con las que ha estado expuesto, si sabe cuales son: ________________________

   ____

   ____

3. ¿Ha trabajado con los siguientes materiales o las condiciones anotadas abajo?:

   a. Asbestos: Sí o No
   b. Sílice (Limpiar mediante un chorro de arena): Sí o No
   c. Tungsteno/Cobalto (pulverizar o soldadura): Sí o No
   d. Berilio: Sí o No
   e. Aluminio: Sí o No
   f. Carbón de piedra (minando): Sí o No
   g. Hierro: Sí o No
   h. Estaño: Sí o No
   i. Ambiente polvoriento: Sí o No
   j. Otra exposicion peligrosa: Sí o No

   Describa las exposiciones peligrosas:

   ____

   ____

4. ¿Tiene usted otro trabajo o un negocio aparte de este?

   ____

   ____

5. Apunte su previos trabajos: ________________________

   ____

6. Apunte sus pasatiempos: ________________________________

   ____

   ____

7. ¿Tiene servicio militar? Sí o No

   Si la respuesta es "Sí", ha estado expuesto a agentes químicos o biologicos durante entrenamiento ocombate: Sí o No

8. ¿Alguna vez ha trabajado en un equipo de HAZMAT (equipo respondedor a incidentes de materialespeligosos con emergencia)? Sí o No

9. ¿Esta tomando alguna medicina que no haya mencionado en este cuestionario (incluyendo remedios caseros omedicinas que compra sin receta)? Sí o No

   Si la respuesta es "Sí", cuales son: ____________________

10. ¿Va a usar algunas de las siguientes partes con su respirador?
    a. filtros HEPA (filtro de alta eficiencia que remueve partículas tóxicas en la atmósfera): Sí o No
    b. Canastillo (por ejemplo, máscara para gas): Sí o No
    c. Cartuchos: Sí o No

11. ¿Cuántas veces espera usar un respirador?
    a. Para salir de peligro solamente (no rescates): Sí o No
    b. Recates de emergencia solamente: Sí o No
    c. Menos de 5 horas por semana: Sí o No
    d. Menos de 2 horas por día: Sí o No
    e. 2 a 4 horas por día: Sí o No
    f. Mas de 4 horas por día: Sí o No

12. ¿Durante el tiempo de usar el respirador, su trabajo es...?
    a. Ligero (menos de 200 kcal por hora): Sí o No

    Si la respuesta es "sí", cuanto tiempo dura la obra

    __________ horas __________ minutos

    Ejemplos de trabajos ligeros: estar sentado escribiendo, escribiendo a máquina, diseñando, trabajando la línea de montaje, o estar parado gobernando un taladro o máquinas:

    b. Moderado (200-350 kcal por hora): Sí o No

Si la respuesta es "sí"cuanto tiempo dura en promedio por jornada

_________ horas _________ minutos

Ejemplos de trabajos moderados : sentado clavando o archivando; manejando un camión o autobús en traficopesado; estar de pie taladrando, clavando, trabajando la línea de montaje, o transferiendo una carga (de 35 libras) a la altura de la cintura; caminando sobre tierra plana a 2 millas por hora o bajando a 3 millas por hora; empujando una carretilla con una carga pesada (de 100 libras) sobre terreno plano.

c. Pesado (mas de 350 kcal por hora): Sí o No

Si la respuesta es "sí"cuanto tiempo dura en promedio por jornada

_________ horas _________ minutos

Ejemplos de trabajos pesados: levantando cargas pesadas (mas de 50 libras) desde el piso hasta la altura de la cintura o los hombros; trabajando cargando o descargando; transpalear; estar de pie trabajando de albañil o demenuzando moldes; subiendo a 2 millas por hora; subiendo la escalera con una carga pesada (mas de 50 libras).

13. ¿Va a estar usando ropa o equipo protectivo cuando use el respirador? Sí o No

    Si la respuesta es "sí" describa que va a estar usando

    ____

    ____

14. ¿Va a estar trabajando en condiciones calurosas (temperatura mas de 77°F)? Sí o No

15. ¿Va a estar trabajando en condiciones humedas? Sí o No

16. Describa el tipo de trabajo que va a estar usted haciendo cuando use el respirador.

    ____

    ____

17. Describa cualquier situacion especial o peligrosa que pueda encontrar cuando este usando el respirador (por ejemplo, espacios encerrados, gases que lo puedan matar, etc.)

    ____

    ____

18. Provea la siguiente informacion si la sabe, por cada sustancia tóxica que usted va a estar expuesto cuando este usando el respirador(s):

    Nombre de la primera sustancia tóxica:______________

    Maximo nivel de exposición por jornada de trabajo:______

    Tiempo de exposición por jornada:______________

    Nombre de la segunda sustancia tóxica:______________

    Maximo nivel de exposición por jornada de trabajo:______

    Tiempo de exposición por jornada: ______________

    Nombre de la tercera sustancia tóxica:______________

    Máximo nivel de exposición por jornada de trabajo: ______

    Tiempo de exposición por jornada: ______________

    El nombre de cualquier sustancia tóxica que usted va a estar expuesto cuando este usted usando el respirador

    ____

    ____

    ____

19. Describa alguna responsabilidad especial que usted va a tener cuando usted este usado el respirador(s) que pueda afectar la seguridad o la vida de otros (por ejemplo, rescate, seguridad).

    ____

    ____

Appendix 1D

Appendix D to § 1910.134 (Mandatory): Information for Employees Using Respirators When Not Required Under the Standard

Respirators are an effective method of protection against designated hazards when properly selected and worn. Respirator use is encouraged, even when exposures are below the exposure limit, to provide an additional level of comfort and protection for workers. However, if a respirator is used improperly or not kept clean, the respirator itself can become a hazard to the worker. Sometimes, workers may wear respirators to avoid exposures to hazards, even if the amount of hazardous substance does not exceed the limits set by OSHA standards. If your employer provides respirators for your voluntary use, of if you provide your own respirator, you need to take certain precautions to be sure that the respirator itself does not present a hazard.

You should do the following:

1. Read and heed all instructions provided by the manufacturer on use, maintenance, cleaning and care, and warnings regarding the respirators limitations.

2. Choose respirators certified for use to protect against the contaminant of concern. NIOSH, the National Institute for Occupational Safety and Health of the U.S. Department of Health and Human Services, certifies respirators. A label or statement of certification should appear on the respirator or respirator packaging. It will tell you what the respirator is designed for and how much it will protect you.

3. Do not wear your respirator into atmospheres containing contaminants for which your respirator is not designed to protect against. For example, a respirator designed to filter dust particles will not protect you against gases, vapors, or very small solid particles of fumes or smoke.

4. Keep track of your respirator so that you do not mistakenly use someone else's respirator.

[63 FR 1152, Jan. 8, 1998; 63 FR 20098, April 23, 1998]

Appendix 2

Environmental Conditions and Respirator Use

Company: ______________________________ Date: ______________

Job: ________________________ Dept: ______________________

Exposures	Representative exposure levels	Permissible exposure level or TLV

I. Type and weight of respirator to be used:
- _____ Dust/mist respirator wt:_____
- _____ Air purifying respirator—please circle (full) (half) wt: _____
- _____ Supplied air wt: _____
- _____ SCBA (weighing up to 35 lb) wt: _____
- _____ Tight Fitting _____ Loose fitting

II. Level of work required during respirator use:
- _____ Observation only (0–2 METS)
- _____ Light work (lifting, pushing, pulling 20 lbs) (2–4 METS)
- _____ Moderate work (lifting, pushing, pulling 21–45 lbs) (4–6 METS)
- _____ Heavy work (lifting, pushing, pulling 45 lb) (6–8 METS)
- _____ Emergency rescue type work (presumed to be heavy work) (> 8 METS)

III. Extent of usage
- _____ Daily: Hours per day: _____
- _____ Occasionally: Days per week _____ Hours per day _____
- _____ Rescue use
- _____ Escape use

IV. Special work considerations, please describe:

❑ Heights ❑ Safety glasses
❑ Extreme temperatures ❑ Gloves
❑ Noise > 85 dB ❑ IDLH
❑ Protective clothing ❑ Confined Space
Verbal communication required

Form Completed by: ________________________ Date: ______________

Appendix 3

Physical Examination Form: Respirator Use Qualification

Name: ______________________________ SSN: __________________

Date: ______________________________

Date of birth: ________________ Company: _________________________

Blood pressure: _____ Pulse: _____ Respiration: _____ Ht: _____ Wt: _____

Exam components address only those areas of concern identified in the health history.

	NL	ABNL	N/A	Comments
Facial (note lesions, deformities, facial hair)				
Ears (note perforated TM)				
Eyes (note glasses/contacts)				
Mouth (dentures, deformity)				
Nose (deformity, patency)				
Lungs				
Hear				
Spine (if indicated by history)				
• ROM				
• Deep knee bend squat				
SLR (sitting/lying)				
Rhomberg				
Diagnostic testing				
• Spirometry				
• Chest x-ray PA/lat. (if indicated)				
• ECG (if indicated)				
• Exercise stress test (if indicated)				
• Olfactory test (if indicated)				
• Work trial (if indicated)				

Additional comments: ____

____

____

Physician or LHCP: ____

Date: ______________

Appendix 4

Consent and Release Form

This consent may be used for a number of different employer requested evaluations, e.g. preplacement, medical surveillance, fitness for duty.

Consent and Release

Name: ______________________________

I the above named person gave my consent for a medical evaluation requested by my employer. I understand the purpose of the evaluation is to assess my medical ability to perform my job or work safely and/or to assess whether I have experienced adverse effects from my work.

I authorize ______*clinic name*______ to release to my employer and other identified party(s) the result of the medical evaluation. It is my understanding that this information will be solely for the purpose described above. I am not giving permission for redisclosure.

Employer: ______________________________

Additional Party ______________________________ ____________

Initials

______________________________ ____________

Employee Signature Date

______________________________ ____________

Witness Signature Date

Appendix 5

Spirometry Testing as Part of the Respiratory Certification Process

In contrast to some Occupational Safety and Health Administration (OSHA) health standards, the respirator certification standard does not specify procedures or the qualifications of technicians performing spirometry. Nonetheless, careful attention to equipment calibration, technician training, and procedures determines the utility of spirometry in predicting a worker's ability to tolerate a respirator. Using the American College of Occupational and Environmental Medicine's (ACOEM) and the American Thoracic Society's (ATS) guidelines, this appendix outlines the essentials for collecting valid and reliable spirometric data for use in respirator certification.[1,2] Of course, spirometric results provide data useful for occupational medical purposes other than respirator certification. In addition, more advanced pulmonary function tests may be appropriate for further evaluation of some individuals with pulmonary problems. Readers requiring more detailed information are referred to the Further Reading section at the end of this appendix.

Properly performed spirometry produces three measurements useful to the respiratory certification process:

- *Forced vital capacity (FVC):* A measure of the total volume of air forcefully exhaled during the maneuver
- *Forced expiratory volume in 1 second (FEV_1):* A measure of the amount of air forcefully exhaled during the first second of the maneuver
- *FEV_1/FVC ratio:* A relationship that reveals possible obstructive or restrictive disease

The ATS discourages use of the FEF_{25-75} and other measures of instantaneous flow to diagnose small airways disease.

Spirometry Program Medical Oversight

A physician should provide medical direction for the spirometry program. The physician may provide this oversight in conjunction with a respiratory therapist with specialized knowledge and experience in pulmonary function testing. Activities include, but are not limited to:

1. Selection of a reference equation for predicting normal lung function
2. Selection of spirometry equipment
3. Identification of the tests conducted and under what circumstance
4. Approval of spirometric procedures and interpretation guidelines
5. Creation and oversight of a quality control program[3]

Technician Qualifications

As outlined in the Preamble to the OSHA cotton dust standard (29 CFR 1910.1043):

> The key to reliable pulmonary function testing is the technician's way of guiding the employee through a series of respiratory maneuvers. The most important quality of a pulmonary function technician is the motivation to do the very best test on every employee. The technician must also be able to judge the degree of effort and cooperation of the subject. The test results obtained by a technician who lacks these skills are not only useless, but also convey false information which could be harmful to the employee.

ACOEM, the National Institute of Occupational Safety and Health (NIOSH), the ATS, and the American Association of Occupational Health Nurses (AAOHN) all recommend technician training to ensure accurate pulmonary function testing.

Suggested minimum qualifications are

1. Successful completion of a NIOSH-approved spirometry course.
2. Six-month supervised training period for technicians conducting spirometry. For technicians whose job description requires problem solving, a 1-year training program is recommended.

3. Periodic intensive refresher courses for experienced technicians every 2 to 3 years.[1]

Quality Program

A spirometry quality program encompasses both quality control and quality improvement. The quality control program concerns characteristics of equipment operation and criteria for the acceptability and reliability of test results. Quality improvement addresses technician skill development and overall program attributes such as communication.

Quality Control

Characteristics of Equipment Operation

Test validity and reliability demands that spirometry equipment work properly. Technicians should maintain a log book of all quality control checks. The log book also should include the date and type of all repair(s) and any software and hardware updates. *Manufacturers of spirometry equipment provide detailed calibration requirements specific to their spirometer. Readers who use this appendix for their spirometry programs should include these details in their clinic manuals.*

ATS Guidelines for Daily Calibration[1]

1. *Volume test.* This test validates the accuracy of a spirometer's volume tests. Conducted at the start of each workday, the test requires that a minimum of 3 liters of air be injected into the spirometer with a calibrated syringe. The syringe must have an accuracy of 0.5 percent of its full volume (15 ml for a 3-liter syringe). To control the variables of temperature and humidity on test equipment, the calibration syringe and spirometer should be stored in close proximity. Volume testing includes three timed tests of air injection: 1 second, 6 seconds, and between 2 and 6 seconds.

When large numbers of spirograms are conducted daily, technicians should perform the volume calibration every 4 hours. When performing spirometry in environments where the temperature fluctuates, technicians also should monitor ambient air temperature and enter changes into the spirometer's database.

Test results: The spirometer should read 3 liters (15 ml) at the end of each test. If volume results do not comply with this standard, the technician should determine why. Meanwhile, the spirometer should not be used. Areas for review include, but are not limited to, spirometer, environment, accuracy of calibration syringe, and tester technique.

2. *Leak test.* A volumetric spirometer requires a daily check to detect air leakage. Consult the manufacturer's guide for conducting a leak test. A leak test is judged unacceptable when the volume over 1 minute varies by more than 10 ml. Technicians should not use spirometers with an unacceptable leak test until the condition is corrected.

3. *BTPS.* All lung volumes are reported at BTPS: at body temperature, ambient barometric pressure, and fully saturated with water. The ATS recommends that technicians should not conduct diagnostic spirometric studies in environments where the ambient air temperature is less than 17°C or more than 40°C. Outside these temperature extremes, potential errors in FVC and FEV_1 may occur. Follow the manufacturer's requirements regarding environmental conditions for testing.

Periodic Calibration

1. *Syringe calibration.* The calibration syringe must be accurate within at least 15 ml or 0.5 percent of its full volume. Consult the syringe manufacturer's recommendations for guidelines regarding the frequency and requirements for syringe calibration. The ATS recommends periodic testing for syringe leaks. This is accomplished by filling the syringe with air, closing off the exhaust valve, and attempting to discharge the air. Discharge of air is considered a positive leak test. In the event of a positive leak test, the syringe is removed from service until repaired.

2. *Linearity test.* The ATS recommends that volume spirometers receive quarterly checks for linearity and that flow-volume spirometers receive a minimum of a weekly check. A linearity test involves injecting the volume of a 3-liter syringe at several different flow rates. An acceptable test is one that meets the volume-accuracy requirement at each of the flow rates.

3. *Time test.* The time test, conducted quarterly, validates the accuracy of the spirometer's internal time recorder by comparing its reports against a stopwatch. The spirometer's reported time must match a stopwatch within 1 percent.

4. *Software.* The ATS's 1994 Spirometry Standardization Guideline reported that during testing of commercial spirometers, a number of errors were identified in computer software systems. In response, the

ATS published computer guidelines and computer software recommendations. With the implementation of new software, the ATS also recommends the installation date be logged and an initial test be conducted on a "known subject." Usually, the technician enlists a colleague as the known subject and compares the subject's previous spirometry test results with those obtained with the new software.

Test Acceptability and Reliability

The ATS identifies reproducibility and acceptability criteria that form the basis of a valid test. Documentation of a quality test statement on the test record by the tester provides the interpreter with meaningful information as to the circumstances and validity of the test.

The ATS spirometry test acceptability criteria require that the technician evaluate the following:

1. *Good start to the test after a maximal inspiration.* Extrapolated volumes of less than 5 percent or 0.15 liter reflect a good start. In cases of a poor start or hesitation, computerized spirometers will print out "extrapolated volume too high."
2. *Satisfactory exhalation:*
    - A minimum 6-second exhalation without early termination of expiration. The technician looks for a plateau in the volume-time curve for a minimum of 1 second. If a plateau does not occur within the recommended 6-second test period, the test is valid if the FVC meets the 1-second plateau requirement. Older individuals or those with a pulmonary obstruction may require longer than 6 seconds to meet plateau requirements, but anything greater than 15 seconds yields little in the clinical decision process. The ATS does not recommend multiple prolonged exhalations. In this case, a slow or unforced vital capacity (VC) may prove more beneficial than a FVC.
    - No cough or glottis closure during the first second of the test. Cough or glottis closure occurring later in the test may or may not adversely affect test results.
    - No observed leaks in the seal of the patient's lips around the mouthpiece.
    - No variability in effort over the course of the expiration.
    - No mouthpiece obstruction by objects such as dentures.

After a minimum of three acceptable tests, the technician evaluates the reproducibility of the results. Poor reproducibility may occur because of poor technician coaching and patient technique or because of the patient's clinical condition. Interpreters should use their clinical judgment about the significance of tests with poor reproducibility. *Unacceptable* tests are not valid tests and are not included in the evaluation of reproducibility. The ATS reproducibility criteria include

1. The two largest FVC values are within 0.2 liter of each other.
2. The two largest FEV_1 values are within 0.2 liter of each other.
3. If the first three tests do not meet these criteria, the technician should repeat trials up to a maximum of either eight tests or the individual's tolerance.

Quality Improvement Process

A continuous quality improvement process for conducting spirometry begins with establishing quality indicators with associated performance expectations. Indicators may address any part of the spirometry program. For example, one indicator may look at required documentation, such as documentation of acceptability criteria. Chart audit results are measured against an established threshold; e.g., 98 percent of all records should show acceptability documentation. An indicator that identifies employee understanding of the test procedure can help ensure completeness of pretest directions. Periodic spirogram review with the technician provides valuable feedback regarding testing technique. The information gleaned from a quality improvement process encourages excellence in practice as well as growth of staff knowledge and skill.

Infection Control

Infection control in the spirometry setting must address two types of infection transmission: direct and indirect contact. Following standard infection-control procedures eliminates infection from direct contact with saliva or blood (mucus membrane trauma from testing). Infection-control practices include

- Technicians wear gloves for removal of mouthpiece or the employee removes and discards the mouthpiece.

- Technicians disinfect, sterilize, or discard reusable equipment that comes in direct contact with a mucus membrane.
- Technicians disinfect tubing with visible condensation.
- Technicians wash hands before and after each employee is tested.

Indirect transmission concerns infections carried by aerosols, such as tuberculosis and viral infections. Following standard infection-control practices minimizes this type of infection transmission. However, the use of a closed-circuit technique may increase the possibility of infection transmission from one patient to the next. In this circumstance, the ATS recommends flushing the system five times over its entire volume range. In-line filters are also available for use but must meet established specifications.[1,4]

Spirometry Test Procedure

The technician's ability to standardize spirometric procedures and properly coach subjects remains the most critical components for achieving valid and reproducible test results. Technicians should follow these steps:

1. Review and document the employee history. The purpose of the prespirometry history is twofold. First, technicians should identify conditions or treatments that may affect test results. Second, they should note any contraindications for testing *in the occupational health setting.* The history should include
    - Current acute illness
    - Smoked a cigarette in the past hour
    - Eaten a heavy meal in the past hour
    - Use of an aerosol bronchodilator in the past hour
    - Use of an oral bronchodilator medication at anytime during the day of the test
    - Suffered from a respiratory infection in the past 3 weeks

(a) *Reasons for postponement of testing:*

- Recent (6 weeks) surgery (cataract, thoracic, or abdominal): Defer from testing until medically approved by treating clinician

- History of pneumothorax: Defer from testing until medically approved by treating clinician
- History of unstable cardiovascular condition (angina, hypertension, recent myocardial infarction, pulmonary embolism): Defer from testing until medically approved by treating clinician
- Thoracic, abdominal, or cerebral aneurysms: Defer from testing until medically approved by treating clinician
- Acute disease that may interfere with testing (nausea, vomiting): Defer until symptoms have resolved

(b) *Reasons for discontinuing test:*

- Syncope after forced expiration
- Chest pain after forced expiration
- Bronchospasm
- Paroxysmal coughing[4]

2. Neither the ATS[1] nor the American Association for Respiratory Care[4] recommends blood pressure screening prior to spirometry. Given the physical stress associated with spirometry and that many workers do not get regular medical care, we believe that the occupational health setting warrants a more conservative approach. We recommend obtaining and documenting a blood pressure reading prior to spirometry testing. Defer testing when readings exceed 180/110 mmHg (severe hypertension), and refer the employee to the primary care provider for urgent evaluation.[5] If lower levels of hypertension are noted, spirometry can proceed, but appropriate follow-up should be arranged.
3. Obtain age, height (without shoes), and weight, and enter values into the spirometer. When appropriate, enter race adjustment into the spirometer. Future prediction equations are likely to improve accuracy related to demographic considerations.
4. Instruct the employee in the spirometry procedure and demonstrate.
   - Review the purpose of the test—assessment of maximum exhaled volume and the amount of air exhaled in the first second.
   - Note that a minimum of three valid trials will be conducted.

In the event criteria for validity are not met, state that no more than eight trials will be conducted.

- Review posture during the test. Head and neck should be held in a neutral position. Avoid bending at the waist to squeeze out more air.
- Ask subject to inhale deeply to maximum inspiration.
- Coach the subject to position the mouthpiece just past the front teeth and seal the lips tightly around the mouthpiece.
- Instruct the subject to "blast" the inspired air into the spirometer and to continue exhalation until told to stop. Inform the subject that he or she will not feel any resistance while exhaling into the spirometer.
- Ask the subject to report any discomfort or light-headedness.

Special Considerations

Sitting versus Standing

Subject posture while performing spirometric maneuvers can affect both test results and safety. Expiratory volumes tend to be larger in adults who stand during the test procedure. However, the ATS allows either a sitting or a standing position, with the caveat that test documentation includes position. A large belly may restrict the recorded volumes of a sitting subject. Requiring the same position in subsequent testing ensures appropriate interpretation of any change in test results.[1]

Spirometry can cause light-headedness or syncope as the result of hyperventilation. Ensuring the availability of a stable chair just behind a standing subject is an appropriate precaution. The sitting position decreases the likelihood of a work-related injury for both the subject and the technician.

Nose Clips

Forced expiratory volume testing does not require the use of nose clips. At their discretion, technicians may use nose clips if they observe a subject exhale through his or her nose during a test.[1]

Further Reading

American Thoracic Society. Standardization of spirometry: 1994 update. *Am J Respir Crit Care* 1995; 152:1107–1136.

American Association of Critical Care. AARC Clinical practice guideline for spirometry: 1996 update. *Respir Care* 1996; 41:629–636.

American College of Occupational and Environmental Medicine. ACOEM Position Paper: Spirometry in the Occupational Setting. Available at *http://www.acoem.org/paprguid/guides/sppaper/spirom.htm.*

Ruppel GL. *Manual of Pulmonary Function Testing,* 7th ed. St. Louis: Mosby–Year Book, 1997.

References

1. American Thoracic Society. Standardization of spirometry: 1994 update. *Am J Respir Crit Care.* 1995; 152:1107–1136.
2. American College of Occupational and Environmental Medicine. ACOEM Position Paper: Spirometry in the Occupational Setting. Available at *http://www.acoem.org/paprguid/papers/sppaper/spirom.htm.*
3. Mahler D, Horwitz M. The pulmonary function laboratory. In: Bone E (ed.), *Pulmonary and Critical Medicine.* St. Louis: Mosby–Year Book, 1998.
4. American Association of Respiratory Care. AARC clinical practice guideline for spirometry: 1996 update. *Respir Care* 1996; 41:629–636.
5. Treadway KK. Screening for hypertension. In: Goroll AH, May LA, Mulley AG (eds.): *Primary Care Medicine,* 3d ed. New York: Lippincott-Raven, 1995:69–72.

Appendix 6
Olfactory Acuity Testing Procedure

Purpose

Olfactory acuity testing is used to investigate a worker's report of an olfactory deficit.

This guideline describes the traditional clinical method of olfactory screening.[1] We include blank control vials with odorants to improve validity and reliability. Even more accurate is a commercial method that employs the use of an odorant booklet with four "scratch and sniff" test spots.[2] If the test booklet is used, follow the directions and scoring procedures provided by the manufacturer. When a worker fails either one of the screening tests, we recommend use of a more extensive olfaction test to rule out false-positive results.

Equipment

1. Four simple odorants and a control in separate closed, impermeable containers such as a test tubes are used. Refresh odorants weekly. Suggested odorants include vinegar, coffee, cinnamon, and banana oil. The control vial should contain distilled water (some tapwater may present odors)

or

2. A commercial four-test "scratch and sniff" booklet.(Currently, 4-panel and comprehensive 40-panel "scratch and sniff" tests are available from Sensonics, Inc. Telephone: 1-800-547-8838, or accessible at *http://www.smelltest.com.*)

and

3. Olfactory Test Form to document the results.

Traditional Clinical Test Procedure

1. Explain the procedure and its purpose to the subject.
2. Identify the patency of each nostril.
3. Subject compresses one side of the nose and sniffs through the other; repeat on the opposite side. Document if the subject is unable to sniff through one or both nostrils.
   - Document any nasal congestion or septum malformation.
   - . Ask the subject to close his or her eyes.
4. For each test, randomly present the control or odorant.
5. Ask the subject to identify the vial containing the odor. Then ask the subject to specifically name or describe the odor.
6. Document the results on the Olfactory Test Form.
7. Many individuals with normal olfactory functioning will not be able to specifically identify an odor. For purposes of this examination, consistent, correct identification of all vials containing an odorant is sufficient to warrant a passing grade.
8. If the subjects is unable to detect one or more of the four odors due to
   - Acute nasal congestion, reschedule the test after symptoms have resolved;
   - Chronic nasal congestion, refer to the appropriate clinician for further evaluation and consultation regarding respirator use recommendation.
9. If the subject is unable to detect one or more of the four odors due to
   - Possible olfactory dysfunction without obvious congestion, schedule for a comprehensive olfaction test (see "scratch and sniff" tests above), or
   - Refer the subject to the primary care provider for further evaluation, or
   - Consult with an appropriate clinician regarding test results and the respirator use recommendation.

Olfactory Test Form

Date: ____________

Name:______________________________ DOB: ____________

Job: ______________________ Respiratory type: ______________

Hazardous exposures during respirator use:____________________

____

Conditions that may interfere with testing:

______ acute nasal congestion (URI)

______ chronic nasal congestion (allergies)

______ deviated septum

______other:____

Examination of nasal passages:

R: ______ normal ______ abnormal: ____________________

L: ______ normal ______ abnormal: ____________________

Odorant	Odorant detected	Odorant identified	Odorant Not detected

Test results: ____ Identified or detected 4/4. ____ Failed ____ /4.

Plan: ____ Reschedule for repeat test.

____ Refer to clinician prior to further testing.

____ Do comprehensive smell test.

Clinician signature: ______________________ Date: __________

Comprehensive smell test results: ____________________

____

Clinician signature: ______________________ Date: __________

References

1. Bickley L, Hoekelman R. *Bates Guide to Physical Examination and History Taking*, 7th ed. Philadelphia: Lippincott, 1999.
2. Doty RL, Kobal G. Current trends in the measurement of olfactory function. In: Doty RL (ed.), *Handbook of Olfaction and Gustation*. New York: Marcel Dekker, 1995:191–225.

Appendix 7
Work Trial Procedure

Overview

Work trials (WTs) provide a practical test of a person's ability to perform specific job tasks while wearing a respirator. Two general types of work trials can be employed. The Level 1 WT addresses safety issues that may arise as the result of sensory or communication deficiencies. Trials evaluate the impact of diminished hearing, smell, or vision. Conducting a Level 1 WT requires attention to safety concerns, but with clinical input can be designed and monitored by non-clinical personnel. The Level 2 WT explores potential psychophysiologic limitations of the worker, should be conducted in a controlled environment, and usually requires clinical professional monitoring. Tailoring a WT to an individual employee's condition and job tasks requires some procedural innovation, however the clinical measures of an employee's respirator tolerance remain largely the same.

Prior to the WT, a clinician must approve the individual for fit testing and the use of a respirator during the trial. In some cases, advanced medical evaluation may be necessary for this purpose. Physicians and licensed health care professionals (PLHCPs) should never conduct a WT in an immediately dangerous to life and health (IDLH) setting, nor for a worker with known diagnoses or symptoms that could result in a catastrophic event precipitated by the WT demands.

Purpose

1. To assess and document the ability of the employee to tolerate a respirator under specified work conditions.
2. To try specific accommodations for an employee who might otherwise not be able to tolerate a respirator while performing the assigned work tasks.

Procedure

1. Design WT procedure, safety considerations, and assessment parameters.
    - Level 1 WT: With reference to the Environmental Conditions and Respirator Use Form (see Appendix 2) specific for the subject's job, design the trial for implementation in a typical work setting with usual environmental conditions, but with care not to put the subject at risk for injury.
    - Level 2 WT: With reference to the Environmental Conditions and Respirator Use Form (see Appendix 2) specific for the subject's job, plan a Level 2 WT for either the actual work place during "non-exposure times", or for a simulated work environment.
    - Never conduct work trials in an IDLH environment.
2. Explain the purpose of the evaluation and procedure.
3. Obtain informed consent prior to a Level 2 WT.
4. Complete the work trial questionnaire with the employee.
5. The duration of a Level 1 WTs depends on the number of tasks that require assessment. Level 2 WTs last about 30 minutes unless terminated prematurely by the worker or the clinician.
6. For a Level 2 WT, perform the following tasks:
    - Calculate the age-specific maximal heart rate.
    - Calculate the trial-termination heart rate.: For conditioned healthy workers, use the following formula: (0.85 × ([220 – age in years)]. For deconditioned workers, use (0.60 × ([220 – age in years)]. For individuals on beta blockers, rely on other trial- termination criteria.
    - Monitor blood pressure at baseline and every 10 minutes.
    - Monitor pulse and respiration at baseline, every 5 minutes for first 10 minutes, then at 10-minute intervals.
    - Assess employee perceived exertion using the Borg CR-10 perceived exertion scale (see Table 3-8).
    - Monitor employee for worrisome symptoms and appearance: report of chest pain, dizziness, discomfort, desire to terminate test, or observation of unusual diaphoresis, pallor, cold

clammy skin, dyspnea, disequilibrium, or facial expression denoting anxiety.

7. Stop the trial if any of the following occur:
   - Pulse greater than 85% of maximal heart rate (60% for deconditioned workers)
   - Blood pressure equal to or greater than 250/120 mmHg
   - Systolic blood pressure drops from baseline
   - Onset of irregular heart beat
   - Respirations > 30 breaths/minute
   - Worrisome symptoms or appearance
   - Employee request to terminate trial
8. Document the results on the Work Trial Form.
9. Review the results with the appropriate clinician for respirator recommendation.

Level 1 Respirator Work Trial

Name: ____________________ Date: ________ Job: ____________

Company: __________________ Department: __________________

Reason for work trial:

____

____

____

What special conditions or accommodations exist during work trial:

____

____

Auditory evaluation:

_____ Meets all test requirements.

_____ Does not meet all test requirements.

Identify test areas not met:

____

____

(attach procedure)

Visual evaluation:

_____ Meets all test requirements.

_____ Does not meet all test requirements.

Identify test areas not met:

____

____

(attach procedure)

Smell evaluation:

_____ Meets all test requirements.

_____Does not meet all test requirements.

Identify test areas not met:

____

____

(*attach procedure*)

Level 1 Work Trial: Sample Procedure for Impaired Hearing

Purpose: Evaluation of hearing capability under actual work conditions (includes all required PPE).

Department: ______________________ Work Area: ______________

Respirator type: ____

Additional PPE: ____

Safety considerations:

1. Do not conduct trial in IDLH environment.
2. Do not leave employee unattended during test.
3. Review relevant workplace safety procedures prior to trial.
4. Review emergency procedures with employee, including alternate communication method such as a hand signals.
5. Discontinue trial if safety instructions cannot or are not being followed.

Level 1 Work Trial: Sample Procedure (continued)

Use actual work areas where required communication (voice or sound) takes place unless employee's inability to communicate in this area presents a safety risk. Test communication should take place at usual distances and body positions (see below).

Test Area	Communication distance	Body position	Type of communication	Pass/fail	Comments

Test area: Actual work area.

Communication distance: Usual distance of worker from communication source. Communication source includes worker to worker, supervisor to worker, alarm signal.

Body position: Position of the employee relative to the communication source, e.g., face to face, parallel, back to face, above to below, or vice versa. Note the position of the employee first.

Type of communication: Voice, alarm, or machinery. To more accurately assess ability, vary verbal communication with each test area.

Pass/fail: Examiner grades whether employee heard communication by observing whether subject was able to demonstrate or repeat what was said.

Comments: Examiner documents any other information useful to assessing whether employee will be able to effectively communicate in the assigned work areas.

Additional comments:

____

____

____

Employee name: ______________________

Test monitor: ______________________ Date: ______________

Level 1 Work Trial: Test Areas Sample Documentation

Use actual work areas where required communication (voice or sound) takes place unless employee's inability to communicate in this area presents a safety risk. Test communication should take place at usual distances and body positions (see below).

Test Area	Commun-ication distance	Body position	Type of commun-ication	Pass/ fail	Comments
Assembly line two	3 feet	Side to side	Verbal	Pass	
Paint booth	6 feet	Back to front	Verbal emergency tone	Fail Pass	Able to hear spoken word at 2 feet

Test area: Actual work area.

Communication distance: Usual distance of worker from communication source. Communication source includes worker to worker, supervisor to worker, alarm signal.

Body position: Position of the employee relative to the communication source, e.g., face to face, parallel, back to face, above to below, or vice versa. Note the position of the employee first.

Type of communication: Voice, alarm, or machinery. To more accurately assess ability, vary verbal communication with each test area.

Pass/fail: Examiner grades whether employee heard communication by observing whether subject was able to demonstrate or repeat what was said.

Comments: Examiner documents any other information useful to assessing whether employee will be able to effectively communicate in the assigned work areas.

Additional comments:

____

____

____

Employee name: ______________________

Test monitor: ______________________ Date: ______________

Level 2 Respirator Work Trial

Name: ______________________ Date: ________ Job: ______________

Company: ____________________ Department: ____________________

Reason for work trial (include pertinent health history):

____

____

____

Current symptoms:

___ URI ___ Asthma ___ Bronchitis___ Shortness of breath

___ Chest pain ___ Palpitations___ Anxiety ___Ear infection

___ Other:

Current medications: ____________________________________

Baseline contraindications to conducting a work trial:

___ Chest pain ___ Shortness of breath ___ Palpitations

___ High anxiety ___ Blood pressure > 180/110 mmHg

___ Other acute medical conditions likely to endanger or invalidate the work trial.

What special conditions or accommodations exist during work trial:

____

____

Level 2 Work Trial

Maximal heart rate (220 – age) = ______
85% (60%) of maximal heart rate = ______

Time	Blood pressure	Pulse (Note any irregu-larities	Respira-tion	Borg CR-10	General Condition
Before trial					
At 5 min					
At 10 min					
At 20 min					
At 30 min					

______ Completed work trial without problems.

______ Was able to complete work trial, but expressed complaints:

____

____

______ Was unable to complete work trial due to: ______________

____

____

____

Signature of trial observer:______________________ Date:__________

Clinician Assessment:

____

____

____

Clinician signature: ______________________ Date: __________

Level 2 Work Trial: Sample Environmental Conditions and Respirator Use

Company: ___Spectacular, Inc.___ Date: ___3/18/00___

Job : ___Sander___ Department: ___Paint___

Exposures	Representative exposure levels	Permissible exposure level or TLV
Lead dust	100 μg/m³	50 μg/m³

1. Type and weight of respirator to be used:
   ____ Dust/mist respirator Wt:
   __x__ Air -purifying respirator
   Please check: __✓__ Full ____Half Wt: __6 lb__
   ____ Supplied air (airline, hood, and suit) Wt: ____
   ____ SCBA (weighing up to 35 lb) Wt: ____
   __x__ Tight fitting
   ____ Loose fitting

2. Level of work required during respirator use:
   ____ Observation only (0–2 METs)
   ____ Light work (lifting, pushing, pulling 20 lb) (2–4 METs)
   __x__ Moderate work (lifting, pushing, pulling 21–45 lb) (4–6 METs)
   ____ Heavy work (lifting, pushing, pulling 45 lb.) (6–8 METs)
   ____ Emergency rescue-type work (presumed to be heavy work)
   ____ (> 8 METs)

3. Extent of usage
   __x__ Daily: Hours per day: __4–7__
   ____ Occasionally: Days per week: ____ Hours per day: ____
   ____ Rescue use
   ____ Escape use

4. Special work considerations. Please describe:
   ____ Heights
   __✓__ Extreme temperatures
   __✓__ Noise > 85 dB
   __✓__ Protective clothing
   __✓__ Safety glasses
   ____ Gloves
   ____ IDLH
   ____ Confined space

Form completed by: ________________________

Level 2 Work Trial: Sample Procedure and Test Documentation

Name: *John Brown* Date: 3/18/00 Job: *sander*

Company: *Spectacular, Inc.* Department: *Paint*

Reason for work trial (include pertinent health history):
Evaluation of respiratory tolerance and stamina for respirator use. A 55- year-old man with controlled mild hypertension and a history of shortness of breath while walking up a flight of stairs. He also complains of a nonproductive early morning cough. He is a 1 PPD smoker × 25 years. He is 30+ lb overweight. He leads a sedentary lifestyle and has never worn a respirator in the past. PMD reports no contraindications to conducting work trial. Last visit to PMD was within the past 6 months. He uses a beta blocker for hypertension.

Examination: Blood pressure: *150/86 mmHg* Pulse: *60 and regular*

Respirations: *12* Heart: *Normal* Lungs: *Decreased breaths sounds at bases, scattered inspiratory wheezes*

Extremities: *No clubbing, cyanosis, or edema*

Spirometry: *Moderate obstructive disease, chest x-ray: unremarkable*

Current symptoms: *x* None

___ URI ___ Asthma ___ Bronchitis ___ Shortness of breath

___ Chest pain ___ Palpitations ___ Anxiety ___ Ear infection

___ Other: ___

Current medications: *Beta-blocker*

Baseline contraindications to conducting a work trial: *None*

___ Chest pain ___ Shortness of breath ___ Palpitations

___ High anxiety ___ Blood pressure > 180/110 mmHg

___ Other acute medical conditions likely to endanger or invalidate the work trial.

What special conditions or accommodations exist during work trial:
None

Level 2 Work Trial: Sample

Safety considerations:

1. Test conducted in nonexposure environment.
2. Employee not left unattended during test.
3. Workplace safety procedures reviewed.
4. Discontinue test with any report of respiratory intolerance.

Test description: Wearing a half-face, tight-fitting, negative-pressure, air-purifying respirator, the employee will: lift a 10-lb part out of a waist-height box; while standing or sitting, simulate removing lead paint from the part (which has already been cleaned of lead paint) with an 8-lb. power sander (each part takes 5 minutes); place the finished part in a box located at waist height; carry a full box of completed parts to storage cart 15 feet from work table (box weighs 40 lbs.)

Maximal heart rate (220 – age) = 165

60% of maximal heart rate = 99 (note on beta blocker)

Time	Blood pressure	Pulse (Note any irregu-larities	Respira-tion	Borg CR-10	General condition
Before trial	148/88	60 and regular	12		Good; no physical or medical complaints. Color good, lungs decreased breath sounds, scattered wheezes
At 5 min		70 and regular	16	3	Color good, speaking in full sentences, no dyspnea
At 10 min	190/96	74 and regular	18	4	No respiratory distress, keeping pace with required work load
At 20 min	185/100	80 and regular	20	5	No respiratory distress, notes a mild feeling of fatigue
At 30 min	190/105	94 and regular	22	7	Feels "winded"—like he just walked several flights of stairs. Color good, no diaphoresis. Lungs: decreased breath sounds, scattered wheezes

_____ Completed work trial without problems.

__*x*__ Was able to complete work trial , but expressed complaints: *Mildly short of breath, "winded" and mild fatigue. Recovery to baseline within 15 minutes. Heart rate limited by medication.*

_____ Was unable to complete work trial due to:

____

____

Signature of trial observer: ______________________ Date: __________

Clinician Assessment

Mild hypertension and mild COPD with possible reversible component. May be aggravated by beta blocker. Physically deconditioned. Would probably not tolerate job as described without conditioning. May benefit from medical intervention.

1. Recommend referral to primary physician to optimize medical management.

2. Recommend period of acclimatization to respirator use and job demands, e.g., 1 hour on, 1 hour off with increase of 1 hour over successive days.

3. Suggest reevaluation by physician or LHCP when up to 8 hours to review tolerance or earlier if not tolerating graduated increase of hours. Meanwhile both employee and supervisor should report any health concerns with respirator use.

4. Consider use of PAPR.

Clinician signature: __________________________ Date: __________

Hood, OSHA definition of, 221
Hypertension
  case study of, 196–198, 200
  and respirator approval, 152–153
  and respirator use, 141–142
Hypertrophic cardiomyopathy, 137
  and respirator approval, 153
Hyposmia, 112

Immediately dangerous to life or health (IDLH) situations
  OSHA definition of, 221, 225
  OSHA procedures for, 231–232
  respirator for, 224–225
Imminence, defined by ADA, 79
Immunologic disorders
  accommodation for, 106
  assessment of, 103–104
  education regarding, 10
  and respirator approval, 104–105
  and respirator use, 102
Inspection, of respirators, 234–235
Interior structural firefighting
  OSHA definition of, 221
  OSHA procedures for, 232–233
Ischemic heart disease, 136–139
  and respirator approval, 149–150
Isocyanate asthma, 2
Isolation, caused by respirator, 17

Job analysis, 33–35

Likelihood, defined by ADA, 79
Long QT syndrome, 137
Loose-fitting facepiece, OSHA definition of, 221
Lung cancer
  accommodation for, 131
  education regarding, 131
  and respirator approval, 130–131
  and respirator use, 118

Medical evaluation, 4–5
  administration and review of exam, 53–55
  approach to, 53–55
  case studies of, 196–210
  choosing physician or health care provider for, 37
  communication in, 37–39, 193–211
  confidentiality of, 39, 193
  consent and release form, 289
  distinguished from respiratory surveillance, 3
  flow chart for, 50
  follow-up, 55–56, 210
  frequency of, 59
  issues determined by, 75–76
  OSHA questionnaire for, 51–53, 266–283
  OSHA requirements for, 36–37, 226–229
  predictive value of, 76–77
  procedures for, 49–71
  qualification form, 281
  quality assurance for, 213–217
  recommendations based on, 193–211
  recordkeeping for, 239
  reevaluation, 56–59
  and respirator approval, 81–83
  supplemental medical history for, 83
  tests performed during, 59–67
Medical records, privacy of, 51
Medically prescribed use, of respirator, 186–188
  approval for, 190–191
  assessment for, 189–190
  case study of, 208, 211
  education regarding, 191
  interview for, 188–189
Medication use
  accommodation for, 157
  assessment of, 155
  education regarding, 157
  and respirator approval, 156–157
  and respirator use, 153–154
Mental clouding, respirator modification for, 80
Metabolic disorders, and respirator approval, 92–102
Metabolic equivalents (METs), 65
MPF (measured protection factor), 15
Multiple sclerosis, 87
Musculoskeletal system, respirator effects on, 20
Myocarditis, and respirator approval, 153

Narcolepsy, 87
Negative pressure facepiece check, 263
Negative-pressure respirators, 11–12
  OSHA definition of, 221
Neurogenic syncope, 87
Neurologic conditions
  accommodation for, 91–92
  assessment of, 88–89
  education regarding, 91
  and respirator approval, 89–91
  and respirator use, 86–88
NIOSH Respirator Decision Logic, 35–36

Obesity
  accommodation for, 100
  assessment of, 96
  education regarding, 100
  and respirator approval, 93, 95. 99–100
Olfactory disorders
  accommodation for, 115
  assessment of, 112–113

Olfactory disorders (continued)
  education regarding, 115
  and respirator approval, 113–115
  and respirator use, 111–112
Olfactory testing, 66–67
  equipment for, 301
  purpose of, 301
  report form for, 303
  traditional procedure for, 302
OSHA Respiratory Protection Standard, 219–240
Oxygen deficient atmosphere, OSHA definition of, 221

PAPRs (powered air-purifying respirators), 9
  advantages of, 12
  APF of, 16
  OSHA definition of, 221
Parkinsonism, 87
Peak exercise capacity, 139
Perceived Rating Exertion Scale, 70
Physician or other licensed health care professional (PLHCP)
  choosing, 37
  communication with, 37–39, 161
  confidentiality standards for, 39
  OSHA definition of, 221
  scope of authority of, 82, 84
Physiologic stress, from respirator, 16–20
Pleuritic pain, and respirator use, 118
Pneumoconioses, 116
  and respirator use, 121–123
Pneumonectomy, and respirator use, 118
Pneumonia
  accommodation for, 128
  education regarding, 127
  and respirator approval, 126–127
  and respirator use, 118
Pneumothorax
  accommodation for, 130
  education regarding, 129–130
  and respirator approval, 128–129
  and respirator use, 118
  risk of, 19
Pocket Smell Test, 66
Positive pressure facepiece check, 263
Positive pressure respirator, OSHA definition of, 221
Powered air-purifying respirators (PAPRs), 9
  advantages of, 12
  APF of, 16
  OSHA definition of, 221
Privacy, 55
  of medical records, 51
Pseudofolliculitis barbae, 207–208, 209
Psychological conditions
  accommodation for, 111
  assessment of, 107–108
  education regarding, 110
  and respirator approval, 108–110
  and respirator use, 106
Pulmonary disorders
  accommodation for, 124–125, 126, 128, 130, 131, 133, 134–135
  assessment of, 120–121
  education regarding, 124, 125, 127, 129–130, 131, 133, 134
  and respirator approval, 121–123, 126–127, 128–129, 132–133, 133–134
  and respirator use, 116–119
  types of, 117

Qualitative fit test (QLFT), 40
  OSHA definition of, 221
  procedure for, 41
  protocols for, 244–254
Quality assurance
  audit tool for, 216
  customer expectations, 214
  identifying areas for improvement, 215–217
  identifying problems, 217
  quality indicators, 214–215
  of spirometry, 293–296
Quantitative fit testing (QNFT), 40
  OSHA definition of, 222
  procedure for, 41–42
  protocols for, 254–262

Recommendation form, for respirator use, 195
Recordkeeping, 45
  OSHA regulations regarding, 239
Reevaluation, 56–59
  triggers for, 57
Respirator failure, 27–28
  reasons for, 28
Respirator intolerance
  accommodation for, 161
  assessment of, 158–159
  education regarding, 160
  and respirator approval, 159–160
  and respirator use, 158
Respirator use
  ANSI protocol for, vii
  environmental conditions and, 285
  improper, 77
  limitations on, 76
  medical evaluation for, 81–84
  medical guidelines for, 84–177
  OSHA standards for, vi–viii, ix, 1, 5, 29–31, 42–43, 230–233
  permissible practice for, 219
  problems with, 185–191
  program evaluation for, 44

qualification form for, 287
record keeping for, 45
voluntary, 186
work trials for, 67–71, 299–301
Respirators
air-purifying, 6, 8–9, 16, 219
atmosphere-supplying, 10–11, 12, 43, 220
characteristics of, 7
choice of, 194
cleaning procedures for, 233, 264–265
employee information for use for, 284
employee training for, 43–44
employer responsibilities regarding, 27–46
failure of, 27–29
fit testing for, 39–42
injuries caused by, 20
inspection of, 234–235
issues regarding, 5–6
job demands and, 21
maintenance and care of, 43
medical prescription of, 186–191
modifications for symptomatic users, 80
pressure dynamics of, 11–13
psychophysiologic demands of, 16–20, 117
recommendation form, 195
repairs of, 235
selection of, 32–36, 224–226
supplied-air, 10–11, 15, 16, 222
tolerance to, 20
types of, 6, 8–14
Respiratory certification
Americans with Disabilities Act and, 78–79
medical basis for, 76
spirometry in, 291–300
validity of, 77
Respiratory disease, prevention of, 2
Respiratory inlet covering, 13–14
OSHA definition of, 222
Respiratory protection program
administrator of, 31
benefits of, 27
elements of, 30, 32
evaluation of, 238
OSHA regulations regarding, 222–224
recordkeeping for, 239
weaknesses of, 28–29
Respiratory surveillance, 1–2, 4
distinguished from medical evaluation, 3
Rib fracture
accommodation for, 133
education regarding, 133
and respirator approval, 132–133
Right ventricular dysplasia, 137
Risk, duration of, 79
SARs (supplied-air respirators), 10–11, 15
APF of, 16
OSHA definition of, 222
SCBAs (self-contained breathing apparatus), 10, 14
Seal check, by user, 263–266
Seizures
accommodation for, 91–92
assessment of, 88–89
education regarding, 91
and respirator approval, 89–91
and respirator use, 86–88
Self-contained breathing apparatus (SCBA), 10, 14
APF of, 16
OSHA definition of, 222
types of, 11
Senses, respirator effects on, 17
Service life, OSHA definition of, 222
Severity, defined by ADA, 79
Skin disorders
case study of, 207–208, 209
respirator modification for, 80
Sleep apnea, 87
Spirometry, 59–60, 116, 117
advantages and disadvantages of, 61, 63
equipment for, 293
equipment calibration for, 293–295
infection control for, 296–297
interpretation of, 61, 62
medical oversight of, 292–293
patient positioning for, 299
quality control of, 293–296
quality improvement process for, 296
reference ranges for, 60–61, 63
acceptability and reliability of, 295–296
technician qualifications for, 292–293
test procedure for, 297–299
Stress, caused by respirator, 16–17
Stroke
and respirator approval, 151–152
and respirator use, 140
Supplied-air respirator (SAR), 10–11, 15
APF of, 16
OSHA definition of, 222

Thoracic surgery
accommodation for, 134–135
education regarding, 134
and respirator approval, 133–134
and respirator use, 118
Tight fitting facepiece, OSHA definition of, 222
Tobacco use
education regarding, 86
and respirator use, 84–86
Transient blindness, 162–163

Tuberculosis
  accommodation for, 128
  education regarding, 127
  and respirator approval, 126–127
  and respirator use, 118

Upper respiratory infections, and respirator use, 118
User seal check
  OSHA definition of, 222
  OSHA-mandated procedures for, 263–266

Valvular heart disease, and respirator approval, 153
Vision problems
  accommodation for, 166
  assessment of, 162–164
  education regarding, 166
  and respirator approval, 164–166
  and respirator use, 161–162
Voluntary use, of respirator, 186–188
  approval for, 190–191
  assessment for, 189–190
  case study of, 208, 211
  defined, 31
  education regarding, 191
  interview for, 188–189

Work trials, 67–68, 305
  clinical monitoring in, 69–70
  for evaluating hearing, 309
  heart rate in, 69–70
  levels of, 68–69
  procedures for, 306–307
  purpose of, 305
  report form for, 308–317
WPF (workplace protection factor), 15–16